T0143405

Educational Data Mining
with R and Rattle

RIVER PUBLISHERS SERIES IN INFORMATION SCIENCE AND TECHNOLOGY
Volume 21

Series Editors

K. C. CHEN
National Taiwan University
Taipei, Taiwan

SANDEEP SHUKLA
Virginia Tech
USA

CHRISTOPHE BOBDA
University of Arkansas
USA

The "River Publishers Series in Information Science and Technology" covers research which ushers the 21st Century into an Internet and multimedia era. Multimedia means the theory and application of filtering, coding, estimating, analyzing, detecting and recognizing, synthesizing, classifying, recording, and reproducing signals by digital and/or analog devices or techniques, while the scope of "signal" includes audio, video, speech, image, musical, multimedia, data/content, geophysical, sonar/radar, bio/medical, sensation, etc. Networking suggests transportation of such multimedia contents among nodes in communication and/or computer networks, to facilitate the ultimate Internet.

Theory, technologies, protocols and standards, applications/services, practice and implementation of wired/wireless networking are all within the scope of this series. Based on network and communication science, we further extend the scope for 21st Century life through the knowledge in robotics, machine learning, embedded systems, cognitive science, pattern recognition, quantum/biological/molecular computation and information processing, biology, ecology, social science and economics, user behaviors and interface, and applications to health and society advance.

Books published in the series include research monographs, edited volumes, handbooks and textbooks. The books provide professionals, researchers, educators, and advanced students in the field with an invaluable insight into the latest research and developments.

Topics covered in the series include, but are by no means restricted to the following:

- Communication/Computer Networking Technologies and Applications
- Queuing Theory
- Optimization
- Operation Research
- Stochastic Processes
- Information Theory
- Multimedia/Speech/Video Processing
- Computation and Information Processing
- Machine Intelligence
- Cognitive Science and Brian Science
- Embedded Systems
- Computer Architectures
- Reconfigurable Computing
- Cyber Security

For a list of other books in this series, visit www.riverpublishers.com

http://riverpublishers.com/series.php?msg=Information_Science_and_Technology

Educational Data Mining
with R and Rattle

R. S. Kamath

Department of Computer Studies,
Chhatrapati Shahu Institute of Business Education
and Research, Kolahpur, Maharashtra

R. K. Kamat

Department of Computer Science,
Shivaji University,
Kolhapur, Maharastra

River Publishers

Published, sold and distributed by:
River Publishers
Niels Jernes Vej 10
9220 Aalborg Ø
Denmark

River Publishers
Lange Geer 44
2611 PW Delft
The Netherlands

Tel.: +45369953197
www.riverpublishers.com

ISBN: 978-87-93379-31-2 (Hardback)
      978-87-93379-30-5 (Ebook)

©2016 River Publishers

Contents

Foreword

"It is a capital mistake to theorize before one has data. Insensibly one begins to twist facts to suit theories, instead of theories to suit facts."

–Arthur Conan Doyle, Sherlock Holmes

Above quote authored way back in 1887 still holds good in the data dominant era of 21st century. Data helps in interfacing the educator with the students, their learning, and, in turn, helps to gain an insight regarding the whole pedagogical process and moreover goads to participate in a dialog with associates, students, and folks. With regards to teaching-learning, I can't help contradicting British physicist Lord Kelvin, who said, "When you can't express it in numbers, your knowledge is of a small and unsatisfactory kind." As it is popularly said "Not everything that can be counted counts, and not everything that counts can be counted" is more apt for the higher education paradigm. In educating, connections and observations matter as much as educational curriculum and methodology. The above truth, in fact, justifies the very rationale of 'Data Mining' which has been put forth throughout the present book.

The metaphor of data, data everywhere, but not a single datum which can be put to real use has motivated the authors to take up research on Educational Data Mining for the analysis and prediction of students' academic performance. The outcome of the research, as presented in this book emphatically presents that, Educational Data Mining is a rising order, worried with creating techniques for investigating the extraordinary sorts of data that originate from instructive settings, and utilizing those strategies to better comprehend understudies, and the settings in which they learn. A paradigm that is propelled to this book is that of learning by example. The expectation is that peruser of this book will have the capacity to effectively imitate the examples given in the book and after that re-tune them to suit their own particular needs.

The fundamental target of the proposed exploration of the book is to apply Data Mining methods to analyze learner's scholarly performance. This exploration presented through six chapters along with a good number of figures, infographics, charts and tables are a contextual investigation of a scholarly establishment to give quality instruction by looking at the data and discover the components that influence understudies' teaching-learning ambience. The principle point of the exploration is to uncover the uses of the high capability of Data Mining for understudies' performance management using the right kind of tools such as 'R' and 'Rattle'.

I am glad for the fact that the pedagogical information, data and realities procured from our institute i.e. Chhatrapati Shahu Institute of Business Education and Research (prominently known as CSIBER), Kolhapur and appropriately researched by the authors is being disseminated as a book at the international level through the River Publishers, Netherlands. CSIBER established by the visionary academician late. Dr. A. D. Shinde, four decades prior, stands for astounding support of the student community and the society at large in the field of instruction and improvement. CSIBER has been known for quality training in the areas of national significance, like management, environment management, social work and computer application and I trust that there will be numerous more quality books to take after as the present one.

Finally, the book has turned out to a rich presentation striving for students and academicians from the higher education background and specialized learning the writers have put over quite a while in this field. I emphatically prescribe this book for budding scholars of Computer Science, academicians in general, administrators in the higher education arena, policy makers and any person who is unequivocally disposed to take up his or her profession in the increasingly important domain of Educational Data Mining.

–Dr. R. A. Shinde

Dr. R. A. Shinde, a well-known academician is Secretary and Managing Trustee of CSIBER, a leading premier institute of higher learning in western Maharashtra, India dedicated to imparting quality professional education and training with the state of art conducive ambience to nurture the learners with apt skill set and the spirit of dignity of the individual, excellence and service. (More about CSIBER at http://www.siberindia.edu.in)

Preface

This book portrays **"Educational Data Mining with R and Rattle."** Nowadays, the biggest challenges, the educational institutions face, are the growth of educational data and the usage of this to improve the quality of decisions to bring quality education. Discovery of knowledge for prediction about students' performance is required in order to achieve this high level of quality in higher education system. Data mining is a powerful technology with great potential that can be used for the extraction of hidden predictive information from large databases. Educational data mining the emerging field concerns with developing techniques that find out knowledge from data initiating from educational environments. It is a promising discipline, concerned with developing techniques for exploring the exclusive types of data that come from educational settings, and making use this useful information for the growth of educational institutes.

The content of this book designed based on the research entitled "**Mining of Educational Data for the Analysis and Prediction of Students' Academic Performance**" carried out by authors. The basic knowledge imparted in this book can be treated as a guide for educational data mining implementation using R and Rattle open source data mining tools. R open source tool is a simple, but very powerful data mining and statistical data processing tool. It has a number of mining procedures, useful for detailed modeling of a dataset. It supports statistical computations and graphical techniques for the ease of visualization. Rattle is a graphical data mining package written in R. It offers GUIs for R, to sophisticated data analyses, and statistical computation. It is a stepping stone toward using R as a programming language, which allows user to rapidly working through the data processing, modeling, and evaluation phases of a data mining project.

An archetype that is motivated to this book is that of learning by example. The intention is that reader of this book will be able to easily replicate the examples given in the book and then re-tune them to suit their own needs. This is one of the underlying principles of Rattle, where all of the R commands that are used under the graphical user interface are also exposed to the user.

This makes a useful teaching tool in learning R for the specific task of data mining. This book is organized in to six main chapters:

Chapter 1: First chapter is devoted to introduce overview of data mining technologies, educational data mining, installation procedures of R and Rattle. The background of research has covered in this chapter. Since this research is focused on mining of educational data, all the related fields are discussed here. It covers motivation for research problem as well as objectives.

Chapter 2: Emerging research directions in educational data mining covered in this chapter. The associated literature survey and prior art is explained here. This chapter reviews the most relevant research carried out in this area to date.

Chapter 3: This chapter is devoted for design aspects and developmental framework of the system. It covers research design and methodology adopted to carry out EDM study and analysis by focusing on data mining process, EDM phases, exploratory data analysis, interactive graphics, etc. Functions and packages of R for the purpose of mining are presented here.

Chapter 4: The objective of this chapter is to provide a framework for model development and hence building classifiers. Classification is one of the most regularly studied and utilized method by data mining and machine-learning researchers. This chapter focused on different data mining algorithms for classifying students based on their academic performance in semester end examination.

Chapter 5: Educational data analysis using clustering approach and design of prominent clusters have been explained in this chapter. Clustering is finding groups of data items such that the objects in one cluster will be similar to each other and different from the objects in another group. Theoretical aspects and implementation of various clustering techniques are illustrated here.

Chapter 6: Further result evaluation, knowledge presentation, results of performance prediction and students segmentation using clustering are covered in this chapter. It enlightens future research directions too.

R. S. Kamath
R. K. Kamat

Acknowledgment

We wish to express our appreciation to the numerous individuals who saw us through this book; to every one of the individuals who gave support, talked things over, read, composed, offered remarks, permitted us to cite their comments and helped with the proof reading, editing and putting the crux of the matter systematically.

We would like to thank Mark de Jongh, River Publisher, Netherlands for enabling us to publish this book. Thanks a bunch to the anonymous reviewers for their critical review of the book proposal and manuscript. We would like to thank Junko Nakajima from River Publishers for effective copy editing and proof reading.

This research would not have culminated into the present book without the help and support of Dr. R. A. Shinde, Managing trustee, Dr. M. M. Ali, Director, Dr. V. M. Hilage, Advisor, Dr. R. V. Kulkarni, Head, CSIBER, Kolhapur. We are also deeply thankful to Prof. (Dr.) DevanandShinde, Vice Chancellor, Shivaji University, Kolhapur, and all faculty members both from Department of Computer Studies, CSIBER and Department of Computer Science, Shivaji University, Kolhapur for their spirited encouragement.

The first author specially wants to acknowledge with deep gratitude guidance received from her teachers who shaped her career. She would also like to express her heartfelt thanks to her in-laws for their moral support and inspiration throughout the writing of this book. She would also like to express her deep appreciation to her husband Mr. Sudhir Kamath for his encouragement. She wish to extend thanks to her daughter Eesha for her understanding and patience during the course of the present work.The second author wishes to thank his wife Rucha and daughter Reva. Without their support and patience this book wouldn't have seen the light of the day.

Last and not least: we beg forgiveness of all those who have been with us over the course of the years and whose names we have failed to mention.

Dr. R. S. Kamath
Dr. R. K. Kamat

List of Figures

List of Tables

List of Abbreviations

DM   Data Mining
EDM   Educational Data Mining
CRAN   Comprehensive R Archive Network
ITS   Intelligent Tutoring System
ID3   Iterative Dichotomiser
CSV   Comma Separated Values
TXT   Tab separated data
ARFF   Attribute Relation File Format
ANN   Artificial Neural Network

1

Introduction

1.1 Introduction

Higher education throughout the world is delivering through universities, colleges, and other recognized academic institutes. To date, these educational institutes are placed in a high competitive environment and are intending to get more advantages. These organizations need sufficient knowledge for a better assessment, evaluation, planning, and decision-making to remain competitive among educational field. Nowadays, the biggest challenges, the educational institutions face, are the growth of educational data and the usage of this to improve the quality of decisions to bring quality education.

Providing quality education to students is a goal of higher education institutions. Discovery of knowledge for prediction about students' performance is required in order to achieve this high level of quality in higher education system. The data mining (DM) techniques are used to extract hidden knowledge among the educational dataset. Data mining is a powerful technology with great potential that can be used for the extraction of hidden predictive information from large databases. Data mining has its impact in various commercial applications including retail sales, remote sensing, bioinformatics, e-commerce, etc. It integrates a large number of techniques from a variety of areas including databases, statistics, machine learning, data visualization, and others. This technology can discover the hidden patterns, associations from educational data.

Data mining technology is widely used in educational field to retrieve useful information from students' data. The concealed samples that are discovered can be used to analyze and resolve the problem arise in the educational field. The use of data mining to identify factors determining students' academic performance has increased greatly in recent years. The data mining prediction has allowed a decision-making tool which can assist better utilization of recourses in terms of students' academic performance. This prediction helps in improving students' performance. This research evaluates

1

students' performance using classification technique. Educational data mining is to extract the hidden knowledge from data repositories used for decision making in educational system. It promotes diverse tools and algorithms for explore the data patterns. Authors present Educational Data Mining with R and Rattle in this book. It mainly personifies data mining process for predicting students' academic performance using the R\Rattle environment and selected R packages for computing.

In real world, predicting the performance of the students is a challenging task. The primary goals of data mining in practice tend to be prediction and description. This book justifies the potentials of data mining techniques in perspective of higher education by presenting a data mining model. It is a case study of an academic institute that tries to improve the quality of education by exploring the data and find out the factors that affect the students' academic performance in semester end examinations. It reports the classification and clustering tasks that are employed to evaluate students' performance. This extracted knowledge that describes students' performance helps earlier in making out the students who need special coaching and allow the teacher to provide suitable advising.

1.2 Data Mining

Data mining is the method of obtaining hidden, unknown, and probably significant knowledge from large amount of data. It uses an amalgamation of a large knowledge base, advanced analytical skills, and domain knowledge to disclose hidden trends and patterns which can be useful in almost any area from business to medicine, etc. It is a knowledge discovery process by studying the large volumes of data from various outlooks and shortening it into useful information; it has become an essential component in various fields of life. It helps in identifying hidden patterns in a large dataset. Data mining uses a following combination to uncover hidden trends and patterns

- Explicit knowledge base
- Sophisticated analytical skills
- Domain knowledge

Without the help of data mining tools and techniques, it is very difficult for any organization to take out hidden patterns from the large databases and data warehouses. Data mining is a process which comprises of extracting interesting novel and useful information from data. Businesses, scientists, and governments use data mining for many years for the study of data like airline passenger records, census data, and the supermarket scanner. Various data mining techniques such as classification, clustering, association rules, regression, genetic algorithm, neural networks, decision trees, nearest neighbor method, etc., can be used to detect the relevant patterns from databases to look into current and past data which can be studied to predict future trends.

As data mining is the art of data analysis, it discovers meaningful insights from data. It is a process of building models for classification, prediction, clustering, etc. The discovery of new knowledge and the building of prediction models assist the data analyst in decision-making process. Data mining has its application in most of the areas like business, government, financial services, biology, medicine, risk and intelligence, science, and engineering. A key factor for success is retrieving information from data and then turning that information into knowledge. Thus, data mining is being applied and feeding new knowledge into human endeavor.

Data mining deals with building models from the collected data. Modeling is the process of transferring data into some structured form in some useful way. The aim is to explore our data, often to address a specific problem, by modeling the world. These build models to gain insights into the real world and so we can predict how things behave. To accomplish this, different data analysis and model building techniques can be deployed. Today, data mining is a collaborative discipline that combines refined skills in computer science, machine learning, and statistics. Data mining techniques are mainly classified as predictive and descriptive techniques as shown in Figure 1.1.

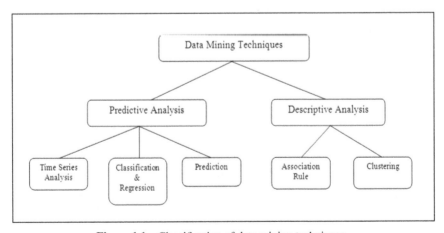

Figure 1.1 Classification of data mining techniques.

DM technology pertain different methods in order to find out and extract patterns from stored data. These trends and patterns form the basis of predictive models that enable analysts to produce new observations from existing data. The pattern extracted can be used to resolve problems of many fields such as business, statistics, education, economic, sport, and medicine. Since the resulting analysis is much more accurate and precise, the huge volume of data stored in those fields demands for data mining approach.

1.2.1 System Architecture

Architecture of data mining system is shown in Figure 1.2. Data repositories for data mining system are database, data warehouse, WWW, or any other repository. Cleaning of data and data integration techniques are applied on the data. It is the process of studying data from different viewpoints and summarizing it into important information so as to identify hidden patterns from a large dataset.

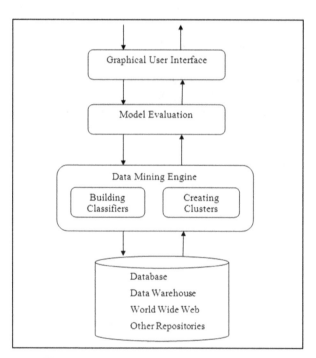

Figure 1.2   Architecture of data mining system.

Knowledge can be extracted with data mining techniques and transforming the databases tasks from storing and retrieval to learning. The essential part of data mining system is DM engine which consists of set of functional module for tasks association, correlation, classification, prediction, clustering, outlier analysis; is and ideally. Model evaluation module is a component that normally includes important measures and search toward interesting pattern by interacting with the data mining modules. Graphical user interface links between the user and the data mining system, allowing the user to communicate with the system by specifying a query or task, providing information to help focus the search, and performing the exploratory mining based on the intermediate results.

1.2.2 Mining Process

The research project documented in this book followed the step-by-step process of a data mining. Figure 1.3 shows the workflow of data mining process

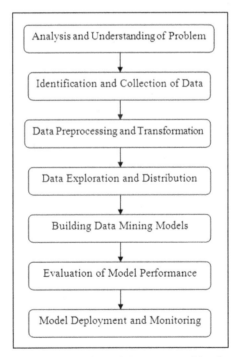

Figure 1.3 Workflow of data mining process with reference to R.

with reference to R. Projects in data mining commonly include following steps:

1. Problem understanding
2. Data collection and understanding
3. Data preparation
4. Modeling
5. Evaluation
6. Deployment

1.2.3 Functions and Products

Data mining function and products include machine learning, statistics, database, and information retrieval as shown in Figure 1.4.

1.2.4 Significance and Applications

Data mining has its impact in various applications; few of them are listed here:

- Banking: prediction of good customers, loan/credit card approval, view the revenue changes by various factors, access statistical information such as maximum, minimum, total, average, trend.
- Retail industry: make out customer buying behaviors, notice customer shopping trends, develop the quality of customer service, attain better customer satisfaction and retention, improve goods consumption ratios, plan more effective goods transportation and distribution policies.
- Telecommunication industry: recognize potentially fraudulent users, find out unusual patterns which may need attention, find usage patterns for

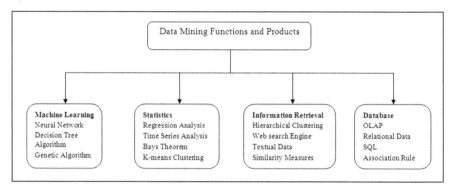

Figure 1.4   Data mining functions and products.

a set of communication services, promote the sales of specific services, and improve the availability of particular services in an area.

- DNA analysis: evaluate the frequently occurring patterns of each class, recognize gene sequence patterns that play roles in various diseases.

1.3 Educational Data Mining—An Area under the Umbrella of Data Mining

The research interests in usage data mining in education is increasing. Educational data mining (EDM), the new emerging field, concerns with developing techniques that find out knowledge from data initiating from educational environments. EDM is a promising discipline which has an imperative impact on predicting students' academic performance. It is a promising discipline, concerned with developing techniques for exploring the exclusive types of data that come from educational settings, and making use this useful information for the growth of educational institutes.

Data mining is an extraction of interesting knowledge from large amount of data. As we know huge amount of data are present in educational database, different data mining techniques are developed and used in order to get required data and to find the unseen relationship. There are varieties of popular EDM tasks, e.g., classification, prediction, association rule, clustering, outlier detection, etc. EDM supports educational system in terms of:

- Predicting student performance relationship between the earlier examination results and their current success
- Forecasting student's academic performance
- Detection of strongly related subjects in the syllabus
- Knowledge discovery on academic achievement
- Classification of students' academic performance according to the learning style
- Finding the similarity and difference between schools.

Educational institutes can make use of data mining to discover precious information from their databases known as educational data mining (EDM). It requires transformation of existing or innovation of new approaches derived from statistics, machine learning, psychometrics, scientific computing, etc.

EDM has emerged as a new research area in recent years and related research areas such as:

- Statistical techniques have been applied to database like student performance, curriculum that was gathered in classroom environments.

- E-learning provides online instruction communication, collaboration, administration, and reporting tools.
- Intelligent tutoring is an alternative to the just-put-it-on-the-web approach by trying to adapt teaching to the needs of each particular student.

Raw data coming from educational systems converted into useful information by EDM process that could potentially have an impact on educational practice and research. It is concerned with developing methods and analyzing educational contents to enable better understanding of students' academic performance. It is also important to improve teaching and learning process. As specified in EDM website http://www.educationaldatamining.org/, EDM is an emerging discipline, concerned with developing techniques for exploring the unique types of data.

EDM provides a set of methods which can help educational organizations to overcome various issues in order to improve learning experience of students as well as growth of institute. Data mining techniques allow user to analyzed data from different dimensions categorized it and summarized the relationship, identified during mining process.

1.3.1 EDM Tasks

Figure 1.5 presents the EDM environment having a few groups of stakeholders who can benefit from EDM in different ways. For instance, learners

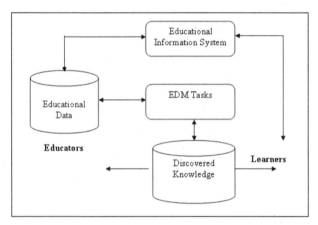

Figure 1.5   EDM tasks.

can receive advice and recommendations about available resources, teachers can see how effective their teaching and what the drawbacks are in the current curriculum. It is expected that the mined knowledge can give a better insight and enhance the educational processes and the learning as a whole.

Basic EDM tasks can be mapped to data mining techniques such as:

1. Classification—categorizing students for determine learning styles and preferences.
2. Predictive modeling—models that can predict performance of a student in semester examination.
3. Clustering—grouping similar students based on academic performance for collaborative learning.
4. Pattern mining—finding patterns including.
5. Visual analytics—reasoning about the educational processes.

1.3.2 Techniques

Educational problems have some special distinctiveness that requires the issue of mining to be treated in a different way. Most of the data mining techniques can be applied directly to handle EDM. DM consists of a set of techniques that can be used to retrieve relevant and interesting knowledge from data. It has several tasks such as classification, prediction, association rule mining, and clustering. Classification techniques classify data item into predefined class label are supervised learning techniques.

In present day's educational system, a students' performance is determined by the internal assessment and end semester examination. The internal assessment is carried out by the teacher based upon students' performance in educational activities such as class test, seminar, assignments, general proficiency, attendance, and lab work. The end semester examination is one that is scored by the student in semester examination.

1.4 Research Problem

Educational data mining is a promising interdisciplinary research area that compacts with the development of techniques to explore data originating in an educational context. It uses computational approaches to study educational data in order to study educational problems. The research problem selected and carried out is stated as **Mining of Educational Data for the Analysis and Prediction of Students' Academic Performance**, and the same is reported here as a book entitled as **Educational Data Mining with R and Rattle**.

1.4.1 Research Motivation

One of the biggest challenges that higher education faces today is predicting the paths of students. The prediction of student's performance is one of the most important wants in order to improve the quality of students. Data mining in educational system is needed for the students as well as academics responsible. EDM promotes the new techniques for retrieving the data that come from educational systems and by using those techniques, prediction can be done for student's academic performance.

The huge amount of data present in educational database that contain useful information aids in predicting of students' performance. Classification is the most useful data mining techniques in educational database. Reported research is aimed at predicting students' performance using EDM.

EDM is a new technique originated from data mining related to field of education. It is the process of transforming the raw data collected by education systems, i.e., exploring hidden data. It promotes distinct techniques and algorithms for the study of data patterns. Here, data are gathered during learning process and then analysis can be done with the techniques from statistics and machine learning. The data mining techniques, like classification, clustering, are applied to extract the hidden knowledge from educational data.

The research here is designed to justify application of various data mining techniques intent in extraction of the hidden knowledge from the student database and prediction of students' performance based on dependency parameters. This aids to identify the students who need special coaching and allow the teacher to provide proper advising. R, open source data mining has been used in this research EDM tool. Various data mining techniques serve this purpose.

1.4.2 Problem Statement

The current educational system is facing numerous issues such as identifying students' requirement, predicting quality of student interactions, etc. Present research intents in extraction of the hidden knowledge from the student database and prediction of students' performance based on performance features such as:

- Previous semester result
- Performance in class test
- Grade scored in assignment
- Attendance
- Performance in practical examination

This aids to identify the students who need special counseling and allow the teacher to provide proper advising. It focused on design of classification model for students' performance prediction by assessing different classifiers experimentally. The objective of the analysis of data is to emphasize constructive information and support decision making. For example, it can help educationalist to analyze the students' learning behaviors and usage information to get a broad view of a students' academic performance.

Figure 1.6 represents the role of educational data mining. The educators design and implement the task to enhance the students' performance. As students are connected with educational system, the dataset can be retrieved from students. These data are given as input to data mining process, and it gives suggestions to students as a result. Various data mining techniques like clustering, classification, and association can be used to extract hidden patterns to the educators.

1.4.3 Objectives

The objective of the reported research is to justify the potentials of data mining algorithms in context of higher education by developing a data mining model. In this research, emphasis is given on extraction of the hidden knowledge from the student database and prediction of students' performance based on dependency parameters. The goal of research is to develop a data mining model with classification and clustering techniques. Classification technique classifies data based on the learning set and applies the model to classify

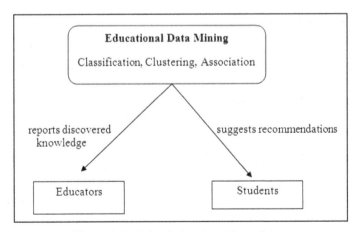

Figure 1.6 Role of educational data mining.

a testing set. Thus, it can be used to evaluate student's performance. This extracted knowledge that describes students' performance helps earlier in identifying. Clustering performs grouping of students based on similarity values. Relationship mining focuses in discovering the relationship between variables.

The objectives of this research are as follows:

- Build the classifier that tests certain features that may affect students' academic performance.
- Forecast of students' performance based on their previous semester result, performance in class test, assignment, and attendance.
- Identify the students who need special counseling by the teacher.
- Understand students' learning processes and to provide feedback to learners.
- Design and analysis of prominent clusters from student dataset.

The main objective of the proposed research is to apply data mining techniques to analyze student's academic performance. This research is a case study of an academic institute to provide quality education by examining the data and find out the factors that affect students' academic performance. It focuses on the implementation of data mining algorithms for acquiring hidden patterns from student dataset. The main aim of the research is to expose the applications of high potential of data mining for students' performance management.

The proposed research intents in find out hidden patterns that could be helpful for the prediction of students' academic performance in end of semester examination based on various dependency parameters.

1.5 R Data Mining Tool

R data mining tool is used throughout this book to illustrate mining procedures [25]. R is a sophisticated statistical and data mining package, easily installed, state-of-the-art, and it is open source. The R commands are saved as scripts, to rerun. These scripts written using language R can accomplish various needs. This book deals with common R functions and commands. For sophisticated and unconstrained data mining, the experienced user will progress to interacting directly with R.

This section is about getting started with data mining in R. We are intent in providing hands-on practice on data mining throughout this book. We have used the open source and free data mining tool Rattle, which is built on the R.

1.5.1 R Installation

In order to install R, a binary distribution from the R Web site (http://www.R-project.org) is obtained. This Webpage contains a link to CRAN (Comprehensive R Archive Network) site (https://cran.r-project.org) to obtain, among other things, the binary distribution for particular operating system/architecture. We have used R-3.1.2, 32 bit version for the experimental purpose. R distribution open source package is downloaded from:

https://cran.r-project.org/bin/windows/base/

This installer contains a binary distribution of R-3.1.2 to execute on Windows XP and later. It is distributed as an installer R-3.1.2-win.exe. R is installed by executing this .exe file. Installation of this creates folder structures as given in Figure 1.7.

To start R in Windows, the appropriate icon R on desktop is simply double clicked. Figure 1.8 shows R console for displaying prompt, indicating R is waiting for commands. The Windows R Console provides set of menus for working with R. These include options for working with script files, managing packages, and obtaining help mainly.

Execution of binary distribution installs some of the packages. Additional packages are installed as per the requirement by selecting CRAN mirror for the session followed by selecting the required package. After connecting computer to the Internet "Install package from CRAN . . ." option in "packages" menu is selected. This option will present a list of the packages available at CRAN. Required package is selected, and R will download the package and self-install it on the system. CRAN is the Comprehensive R Archive Network, provides R binaries, manuals, and packages.

https://cran.r-project.org/

At present, the CRAN package repository contains 7268 available packages. Snapshots of selecting CRAN Mirror and Package from selected mirror are shown in the Figure 1.9. Downloaded package will be unzipped and extracted into library folder of R. To check currently installed packages,

>installed.packages( )

This command displays the list of already installed packages with version details.

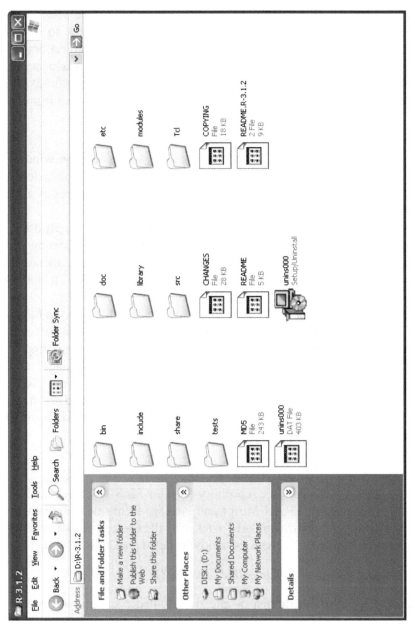

Figure 1.7   Folder structure after installation of R.

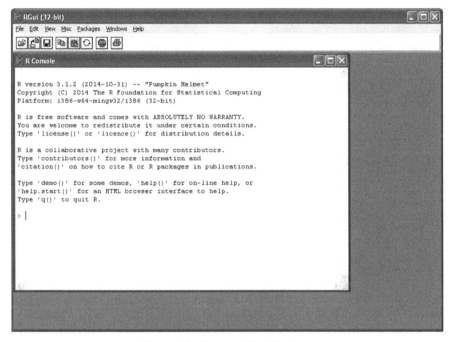

Figure 1.8 R console for Windows.

R has an integrated help system to know more about the system and its functionalities. R comes with a set of HTML files that can be read using a Web browser. Following is the command to HTML help pages.

>*help.start( )*

1.5.2 R Mining

Present research attempts to design data mining model in R, a free software environment. R is a simple, but very powerful data mining and statistical data processing tool for research. R tool is commonly used for statistical computation and data analysis. User friendly and extensibility has achieved R's popularity significantly in recent years. R tool supports statistical computations and graphical techniques for the ease of visualization. Figure 1.10 gives components of R mining.

R is an object-oriented language. All variables, data, functions, etc., are stored in the memory of the computer in the form of named objects. R is a language and an environment for statistical computing.

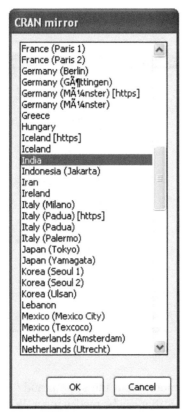 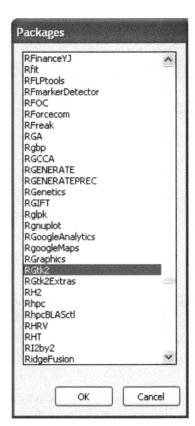

Figure 1.9   Selecting CRAN mirror and package from selected mirror.

There are thousands of contributed packages for R, written by many different authors available for download from CRAN, Comprehensive R Archive Network, http://CRAN.R-project.org and its mirrors. Table 1.1 explains details of R functions and corresponding packages used in this research. R has a number of mining procedures, useful for detailed modeling for a dataset, or for predictive modeling of smaller datasets. Various data mining techniques and algorithms are listed in Figure 1.11.

1.6 Rattle Data Mining Tool

R provides an influential platform for data mining. But scripting and programming is a challenge for data analysts during data mining. Rattle, the ***R Analytical Tool To Learn Easily***, is a graphical data mining package

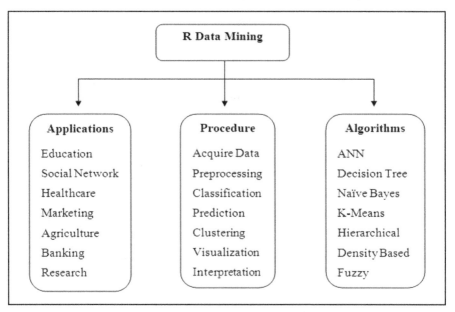

Figure 1.10   Components of R mining.

Table 1.1   R Functions and packages for data mining

	Functions	Library
Classification	Nnet	Nent
	Rpart	Rpart
	Ctree	Party
	naiveBayes	class, e1071
	randomForest	randomForest
	kNN	RWeka
Clustering	agnes	Cluster
	Diana	
	Clara	
	hybridHclust	hybridHclust
	mutualCluster	
	K-means	Stats
	dbscan	Fpc
	cmeans	E107
Association	Elcat	Arules
	NBSelect, NBMiner	arulesNBMiner
	cspade	Arules
		arulesNBMiner

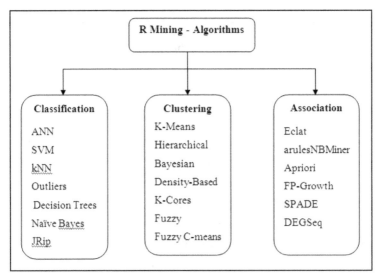

Figure 1.11   R mining—techniques and algorithms.

written in R. It offers GUIs for R, to sophisticated data analyses and statistical computation. The Rattle package provides a graphical GUI for data mining using R. Also it is a stepping stone toward using R as a programming language for data analysis. Rattle is user friendly, allows to rapidly working through the data processing, modeling, and evaluation phases of a data mining project.

In the context of Rattle, the workflow of data mining project can be summarized as:

1. Loading a dataset.
2. Selection of variables for data exploration and mining.
3. Data exploration and understand its distribution.
4. Transformation of data for building data models.
5. Construction of data mining models.
6. Evaluation of these models on other datasets.
7. Exporting the models for final deployment.

Rattle uses the Gnome graphical user interface as provided through the RGtk2 package

1.6.1 Rattle Installation

The following command is entered at the R prompt. R will ask for a CRAN mirror. Nearby location is chosen.

> *install.packages("rattle")*

Rattle can install other packages as needed, but dependent packages can be installed at the time of Rattle installation itself by issuing command:

> *install.packages("rattle", dep=c("Suggests"))*

1.6.2 Loading Rattle Package

Rattle is loaded by giving the following command. Name of the package is supplied as argument to load it.

> *library(rattle)*

After loading Rattle package, then GUI is launched by the command *rattle()*. Corresponding GUI has shown in Figure 1.12.

> *rattle()*

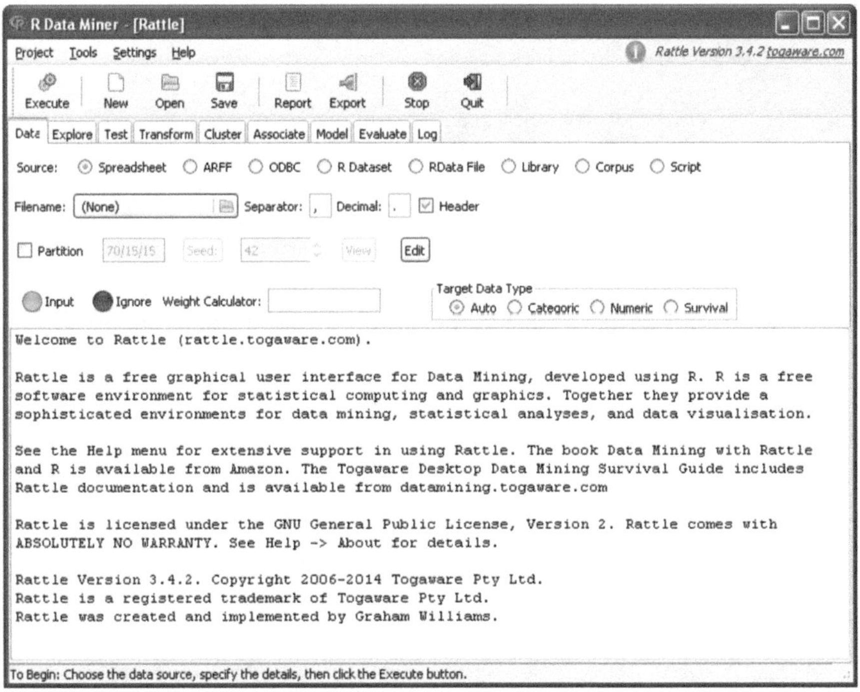

Figure 1.12   Rattle user interface.

Rattle user interface is tab based, sequence of tabs mimicking the typical data mining process.

1.7 Reason for R and Rattle

R and Rattle are free open source software packages with following list of features:

1. It is the most comprehensive statistical package developed for statistical analysis.
2. It provides excellent graphical capability.
3. Since there is no license restriction, it can be run anywhere anytime.
4. R has over 7000 packages available from multiple repositories.
5. It is cross-platform, that is, runs on many operating systems and different hardware.
6. Data can be imported here from Excel, SAS, SPSS, Oracle, etc.

2

Emerging Research Directions in Educational Data Mining

2.1 Introduction

Quite a few studies have used data mining techniques for extracting knowledge and predicting behaviors in the case of healthcare, science, education, information technology, business, human resources, and biology. Educational data mining is focused on developing techniques and studying educational data for the better understanding of students' academic performance. It plays a vital role to enhance teaching and learning methodology. Literature survey reveals that EDM has been an emerging area of research in the current decade. The associated literature survey and prior art is explained in this chapter. This study is supported by referring about 50 research articles. Few of the selected references for broad outline are mentioned here.

2.2 Prior Art Vis-à-vis of Research

In this literature review, references of the similar work have taken and explained the same with respect to this research. This chapter reviews the most relevant research carried out in this area to date.

2.2.1 Educational Data Mining

Guleria and Sood [14] have explained knowledge discovery in data store is the method of finding knowledge in huge amount of data where as data mining is the central part of this task. The whole data mining process is broken into action plan to be executed on data: collection, transformation, mining, and interpretation of results. Authors have studied knowledge discovery aspect in data mining and merged different areas of data mining, algorithms, and methods in it. The study reports techniques like decision tree can predict the class result of students based on the attributes taken. Decision tree classifiers are used on student's data to predict the student's performance in class result. K-means clustering through which student's clusters are formed based on some attributes like their performance, Attendance in class and internals.

Berland et al. [9] have described educational data mining and learning analytics an application to constructionist research. It is a powerful framework for teaching intricate content to beginners. At the core of it is the idea that by enabling students to build innovative artifacts, those learners will have prospects to learn this content in contextualized, meaningful ways. With reference to this researchers investigated the relevance of a set of methods called "educational data mining" to help present a base for quantitative research on constructionist learning. Authors recommend that EDM have the prospective to support research that is useful to researchers working in the constructionist field and also to other communities. They explore collaborations between EDM researchers and constructionist traditions; such partnerships have the potential to develop the skill of constructionist researchers to build rich conclusion about learning, at the same time providing EDM researchers with various interesting new research challenges.

Researchers, Aher and Lobo [19], have surveyed an application of data mining in education system and also presented result analysis using WEKA tool. This paper reports the performance analysis of final year UG Information Technology course students and presented the result using WEKA tool.

2.2.2 Data Mining Using R

Researchers Kumar et al. [13] presented a scalable framework intended at providing a platform for designing and using data mining applications on heterogeneous areas. This framework includes a library of high-performance kernels and a software infrastructure. In addition, it includes a range of optimizations that helps in increasing the applications' throughput. The framework extents multiple technologies comprising R, GPUs, multi-core CPUs, MPI, and parallel-netCDF for high-performance computations. This research framework provides an user friendly and scalable environment both for application development and execution. This framework is exists as a software package that can be easily included in the R programming environment.

2.2.3 Mining Students' Academic Performance

Authors from Kampus Gong Badak have designed a framework for predicting the academic performance of computer science bachelor students. This work reviews the elements required to make prediction of students' performances which are variables, techniques, and tools. Students' details at Faculty of Informatics, University Sultan Zainal Abidin, Malaysia is chosen as a case study for predicting students' performances based on the selected parameters. Naive Bayesian Classifiers is applied for pattern extraction using Weka data mining tool. All these mechanism make up the framework for prediction of students' academic performances.

Yet another paper by Boraka and Rajeshwari explains impact on predicting students' academic performance. In this paper, student's performance is evaluated using association rule mining algorithm. Research has been done on assessing student's performance based on various attributes. In this study important rules are generated to measure the correlation among various attributes which will help to improve the student's academic performance. Experiment is taken using Weka and real-time dataset available in the college premises. The main objective, prediction of student's performance in university result on the basis of their performance in Unit test, assignment, graduation percentage, and attendance is achieved in this research.

2.2.4 Factors Affecting on Students' Academic Performance

Pimpa Cheewaprakobkit [4] has done the analysis to recognize the weak students whose academic performance can be improved. In this research, WEKA data mining tool is used to estimate features for student's academic performance prediction. A cross-validation with 10-folds is applied to predict the accuracy. Two classification algorithms Neural Network and C4.5 decision tree have been accepted and distinguished. The research methodology consists of data preprocessing, attribute selection, and building classification model. According to this study, the decision tree is selected as classifier more accurate than the neural network. It has been concluded that the decision tree model has better efficiency for this dataset.

2.2.5 Evaluation of Student Performance

Noah et al. [10] in his paper, presented the performance evaluation of students, using data mining algorithms. This is a case study accessed students details from University of Port-Harcourt. Dataset designed through University Matriculation Examination and Basic studies programme with the goal of finding out variations in students performance when they graduate. Data mining techniques applied to identify the ratio that falls into grouping of the grading in the various categories using the cumulative grade point and the students who are unsuccessful. The research was able to cluster, analyze, and report the relative performance of each of the classes of the students used in this research. This implementation used Apache, MySQL, PHP, NetBeans IDE 6.8, and XAMPP server.

The work in this area is reported by Syeda et al. [20]. They have presented various classification algorithms, which are applied on various dataset to find out the algorithm efficiency and performance improvement. This can be done by applying data preprocessing tasks, attribute selection, and new class labels prediction. WEKA tool is used for the analysis of various classification techniques such as J48, neural net, decision tree.

Yet another paper by Cohen and Beal [12] surveyed a range of data mining algorithms for the study of how students interrelate with intelligent tutoring systems (ITS). It includes methods for handling hidden state parameters, and for testing hypotheses. Authors have drawn data from two ITSs for math instruction for the illustration of this method. Educational datasets present new challenges to the data mining community which consists of inducing action patterns, design of distance metrics, and inferring unobservable states. Researchers presented ITS data that have structure at various scales, from micro-sequences of hints, to small sequences of actions, and to during sessions long-term patterns. They have confirmed the utility of models with hidden state.

2.2.6 Knowledge Management System

Researchers Natek and Zwilling [16] have focused on data mining techniques for student datasets and intents to answer the research questions. For this, authors have used data and comparative analysis using data mining tools normally available to higher education teachers. This study concluded to encourage teachers to include data mining tools as a part of higher education knowledge management systems. This research explores the opportunity to predict the success rate of students using modern data mining tools available to higher education teachers.

Weka data mining tool was used as comparative study to MS Excel showed that by decision tree models obtains high prediction accuracy. This model able to answer research questions like

1. Before or during a course higher education teachers try to approximate the percentage of successful students.
2. Possibility of predicting the success rate of students in the course.
3. Identification of explicit student characteristics linked with the student performance by the teacher.

4. Availability of related student data to teachers on which helps in predicting performance.

2.2.7 Placement Chance Prediction

Authors Ramesh et al. [22] have designed and reported classification model for prediction of students' campus placement. This paper presents the accuracy of data mining techniques in predicting the performance of a student, which is a great apprehension to the managements of higher education. The first step of the study was to gather student's data. Authors have collected dataset of 300 Computer Science Graduate students, from a private Educational Institution. From the existing database information's such as performance in English, Maths, Programming language, Placement Details were retrieved. Data preprocesses and attributes selection done in second step. The classification technique is applied in this research to evaluate student's performance. Classifiers such as NaiveBayesSimple, MultiLayerPerception, SMO, J48, REPTree were constructed, and corresponding performances were evaluated. The analysis revealed that the MultiLayerPerception is more accurate than the other algorithms. This research helped the institute for prediction of students' academic performance.

Ajay Kumar Pal [11] has reported a new approach of classification for students' placement prediction. This approach presents the relations between academic details and placement of students. In this study, various classification techniques are applied by using WEKA, data mining tool for study of student's academic details. Here, the training algorithm used a set of predefined attributes. The commonly used classification algorithms are naïve Bayesian classification algorithm, multilayer perceptron, and C4.5 tree. The naïve Bayesian classification is the best technique for the high-dimensional inputs. For vector attribute values for more than one class, Multilayer perceptron is most suitable. Currently, C4.5 is most commonly used algorithms due its added features.

2.2.8 Mining Association Rules in Student's Data

The volatile growth of educational data and its usage to improve the quality of managerial decisions for quality education is the biggest challenge of the educational institutions. Pioneering work in this field is reported by Kumar et al. [21]. This paper reports a case study of a university that tries to improve the quality of education by studying the data and discover the features that affect the academic results. In this viewpoint, association rules techniques have used to compare the student's performance. Study considered the subjects common at Graduation and Postgraduation level and will predict the features which can explain their success or failure. The discovered association rules expose various factors such as student's interest, curriculum design; teaching and assessment methodologies that can affect students' performance. The importance of data preprocessing in data analysis and its significant impact on the accuracy of the predicted results is explained here.

Jha and Ragha [8] in their paper surveyed the related studies carried out in EDM using Apriori algorithm. This paper points out the main issues on the application Apriori algorithm in EDM and provides an improved support-matrix Apriori algorithm based on the analysis and research. The proposed improved Apriori uses bottom-up technique with standard deviation functional approach to mine frequent educational data pattern. This algorithm replaces arbitrary user specified minimum support with functional based on standard deviation. In this case, minimum support value is calculated on the basis of standard deviation value of support counts. This approach is more comfortable for somebody non-expert in data mining.

2.2.9 Clustering Data Mining

Patel et al. [6] presented comparative study of clustering techniques in their paper. Five types clustering data mining techniques such as partitioning clustering, grid-based clustering, hierarchical clustering, density-based clustering, and model-based clustering are reviewed here. Finding of this study reveals that all methods perform different role depending on type of data assign and type of application. But still from analysis, authors have concluded that K-means method perform better than other method in many domain.

López et al. [23] from Spain proposed classification via clustering to predict student's performance in university examination on the basis of forum data. This approach obtained similar correctness to traditional classification algorithms.

2.2.10 Prediction for Student's Performance Using Classification Method

Abeer et al. [2] performed the classification task for the final year students' performance prediction. The decision tree (ID3) method is applied here for the classification. For this, some attribute was selected from the student's database used for the prediction. This analysis help the student's to improve the student's performance and also to identify those students who needs special coaching to reduce failing ratio and taking suitable action at right time.

Agarwal et al. [17] have described classification and Decision Tree approach for educational data. A student dataset from a community college was taken and classification techniques were applied and a comparative study has been done. Researchers have chosen 8 different classifiers in WEKA for comparative analysis of performance of classifiers. In this research, Support Vector Machine is selected as a best classifier with good accuracy and minimum root mean square error. This analysis includes a comparative study of all Support Vector Machine Kernel types and among these the Radial Basis Kernel is identified as appropriate for Support Vector Machine. This research is intended to develop a faith on data mining algorithms, so that present education can accept this as a strategic management tool.

Authors Bhardwaj and Pal [3] have designed a data mining model for prediction of students' academic performance so as to recognize the disparity between high learners and slow learners. As a result, 300 students' dataset were used for by Byes classification and prediction model construction. The classification task is applied to evaluate student's academic performance. In this analysis, the decision tree method is used. By this task, knowledge is retrieved that explains students' academic performance in semester end examination. It assists earlier in identifying the dropouts and students who need special coaching and allow the teacher to provide proper counseling.

2.2.11 Classification Techniques

Authors Romero et al. [5], have explained data mining algorithms to classify students. Classification is a supervised learning technique that categorize data item into predefined class label. This is the most commonly useful techniques in data mining to construct classification models from given dataset. There are different classification techniques, such as

- Statistical classification in which individual items are placed into groups based on the quantitative details. For example, k nearest neighbor, least mean square quadratic, and linear discriminate analysis.
- A decision tree is a set of rules organized in a hierarchical structure. This is used as a predictive model in which a data item is classified by following the path of satisfied rules from the root of the tree until reaching a leaf, which is a class label. It can be directly transformed into a set of IF-THEN rules a most popular form of knowledge representation. C4.5 and CART are the most well-known decision tree algorithms.

- Rule Induction a machine-learning technique in which IF-THEN production rules are retrieved from a set of observations. This algorithm is considered as a heuristic state-space search. Here, a state represents a candidate rule and operators correspond to operations that transform one candidate rule into another. AprioriC, CN2, XCS, Supervised Inductive Algorithm, genetic algorithm are some of the examples of rule induction algorithms.
- A neural network a computing paradigm consists of interconnected processing elements called nodes that work together to produce an output function. multilayer perceptron, incremental RBFN, decremental RBFN, radial basis function neural network, hybrid Genetic Algorithm Neural Network, and Neural Network Evolutionary Programming are some of the examples of neural network algorithms are

2.2.12 Educational Data Mining Model Using Rattle

Authors Sadiq Hussain and Hazarika [15] have proposed a method using Rattle for the selection of educational data mining model. This study has considered the data from Dibrugarh University Examination Branch for the affiliated colleges of the University B.A. programme from 2010 to 2013. The model evaluates performance gender wise as well as caste wise of the students. The Colleges are categorized as Urban as well as Rural depending on their locations.

Yet another paper by Williams [7], explained about Rattle, A data mining GUI for R. R provides a powerful platform for statistical computing and data mining. Scripting and programming is a challenge for data analysts for data mining. This issue is resolved by Rattle package which provides a graphical user interface for data mining using R. Rattle is a stepping stone toward using R as a programming language. Williams [7] in his paper explained installation and implementation of Rattle, also explained building of data mining models in Rattle. It is a graphical data mining tool written in and providing a pathway into R. It satisfies the need of basic data mining to sophisticated data analysis with the help of a powerful statistical language.

2.3 Conclusion

The literature review reveals that data mining methods are often implemented at educational institutions for analyzing available data and extracting information and knowledge to support decision-making. Analysis will help the teachers to take proper attention towards the student's progress during the course. Thus, it assists in building reputation of institute in the field of education. Education data mining, the objective of proposed research helps in predicting students' performance in order to recommend improvements in academics. The analysis will reveal dependency factors of student's academic performance. Forecasting will help to identify poor students and additional guidance will provide to improve the result.

Present research intents in extraction of the hidden knowledge from the student database and prediction of students' performance based on performance features. This research applied data mining approach to study and analyze student's academic performance. This aids to identify the students who need particular attention and allow the teacher to make available appropriate advising. The goal of research was to develop a data mining model with classification and clustering techniques. Classification technique classifies data based on the learning set and uses the pattern to classify a test set. Clustering performs grouping of students based on similar values.

3

Design Aspects and Developmental Framework of the System

3.1 Introduction

Data mining methodologies used to study students' performance. This research focuses on development of a data mining model with classification and clustering techniques. Classification technique classifies data based on the learning set and uses the pattern to classify a test set. Clustering performs grouping of students based on similarity values. Information such as previous semester result, performance in class test, grade scored in assignment, attendance, and performance in practical examination was collected from the students' management system, for the prediction of academic performance in the end of the semester. This research is benefited by educational data mining in extraction of the hidden knowledge from the student database and prediction of students' performance based on dependency parameters. This helps to identify the students who need particular attention and allow the teacher to make available appropriate advising.

There are various EDM methods and algorithms adopted in this research to find out hidden patterns and their relationships. These EDM methods include prediction, classification, clustering, and relationship mining.

- The objective of prediction was to construct a model which predicts students' academic performance in semester examination.
- Classification is one of the prediction types which classifies data based on the learning set and uses the pattern to categorize a test data.
- Clustering is the method of grouping data items in classes that are similar, and dissimilar to data items in other classes.
- Relationship mining discovers the relationship between variables.

3.2 EDM Phases and Research Framework

Educational data mining is focused on transforming raw data collected from educational system to translation of new hidden information. EDM mainly consist of four phases:

1. The first step is to find the relationships among educational data of the organization. The goal was to utilize these established relationships in various data mining techniques such as classification, regression, clustering.
2. The second phase is the verification of discovered relationships so that over fitting can be stopped.
3. The third phase is to create predictions on the basis of verified relationships in learning environment.
4. The last phase is sustaining decision making process with the help of predictions.

The structure of this chapter follows the design framework of a data mining process as explained in Section 3.5. In this research, a variety of machine-learning techniques have been applied for the analysis of educational data. A step wise procedure followed for the design of research framework comprises:

1. Problem Definition: Mining of Educational Data for the Analysis and Prediction of Students' Academic Performance.
2. This research has accessed the student database of CSIBER, a postgraduate institute. A set of 200 records of Master of Computer Application students has selected.
3. Dataset is preprocessed as per the requirement of machine-learning algorithms.
4. In supervised learning phase, classifier models are developed which associates the class variable and the descriptive variables using a learning set randomly selected from the dataset.
5. Performance of the each classifier evaluated and selects the "best" one. It allows checking the performance of the trained classifiers against a testing set, evaluating their predictive accuracy with data not used in the training step.
6. In unsupervised learning phase, various clustering techniques are applied to get prominent clusters.

3.3 Methods of Educational Data Mining

There are various promoted methods of EDM but all kind of methods belongs to one of following specified groups:

1. Prediction aims at developing a model which can infer a single facet of data from combination of other facets of data. Classification, regression, and density estimation are the three types of prediction. In any of these categories, the input variables are either categorical or continuous. Classification uses the categorical or binary variables, but in regression uses continuous input variables where as density estimation uses various kernel functions.
2. Clustering technique divides the dataset in various groups, known as clusters. The clustering is more useful if dataset is already specified. According to clustering principle, the data point of one cluster is more similar to other data points of same cluster and more dissimilar to data points of other cluster. There are two ways of instigation of clustering algorithm; start the clustering algorithm with no former assumption and second is to begin clustering algorithm with a prior postulate.
3. Relationship mining generally refers to designing new relationships among the variables. It can be applied on a large dataset with number of variables. This type of mining is an attempt to discover the parameter which is closely associated with the specified parameter. Association rule mining, correlation mining, sequential pattern mining, and causal data mining; types of relationship mining.
4. Discovery with models includes the design of model based on prediction, clustering and knowledge engineering etc. Thus, newly created model's predictions are used to find out a new predicted variable.

3.4 Algorithms and Tools

Educational data mining includes various algorithms that are different in their methods and goals. It also comprises exploration of data and visualization to provide results in a suitable way to users. This research is a case study of EDM process for predicting students' academic performance using the R/Rattle data mining tool with selected R packages for computing. This section presents some algorithms and tools that have used such as

- **Tools**: Student dataset required for analysis is collected and stored in the form of Excel spreadsheets. Initially worked with Excel for visualization.

R/Rattle, open source data mining tool is used for clustering, classification, and association rule.

- **Data exploration and visualization**: Raw data and algorithm results can be visualized via tables and graphics as well as symbolic data analysis. The aim was to present data along with attributes and make excessive points, trends, and clusters noticeable to human eye.
- **Clustering techniques** aim at finding consistent groups in data. K-means technique and its blending with hierarchic clustering are used in this research. Both methods rest on a distance concept between individuals, Euclidian distance is followed.
- **Classification** is used to forecast values for some variable. As an example, this research focuses on extraction of the hidden knowledge from the student database and prediction of students' performance based on dependency parameters. To carry out this various classifiers are developed and tested with dataset.

3.5 Data Mining Process

Figure 3.1 depicts educational data mining process adopted in this research for the analysis of students' academic performance. At the beginning of the study, it is required to find out which data are most powerful influencer for final grade prediction. Evaluation is the way of presenting data mining results to the users which are important because the application of the result is dependent on it. A variety of visualization graphical strategies are used at this phase. Knowledge requires different kinds of representation.

3.5.1 Data Collection

The first step of the study is to gather student's data. This includes collecting the data needed for data mining process and can be acquiring from various sometimes heterogeneous data sources. This research has an access to the related databases of CSIBER, a postgraduate institute for the solitary use of analysis and discovery of knowledge using a data mining approach. A set of 200 records related to Master of Computer Application students have collected. The data were collected from examination department and department of Computer studies. The attributes used are presented in Table 3.1 with the values of every attribute. The study revealed and identified the key factors that decide the outcome of students' performance. Here, the outcome is a categorical variable "grade" with six categories.

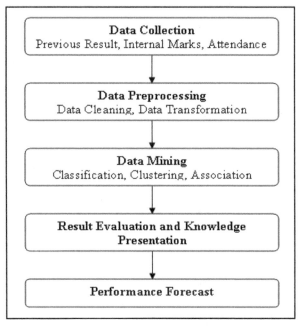

Figure 3.1   Data mining process.

Table 3.1   Features affecting students' academic performance

Variable	Description	Possible Values
Roll	Roll Number	Integer - Continuous
Prev	Previous exam grade	[S : 91–100
		E : 81–90
		O : 71–80
		A : 61–70
		B : 50–60
		X : <50]
attendance	Current attendance	[poor, average, good]
Internal	Performance in internal	[poor, average, good]
Midterm	Midterm exam performance	[0–100]
Laboratory	practical exam performance	[0–50]
Grade	Grade scored current semester	[S, E, O, A, B, X]

This research has collected real-time data that explains the relationships involving learning behavior of students and their academic performance. Partial dataset is illustrated in Figure 3.2. This dataset contain students' details related to academic performance are recorded and subjected to the data mining process.

Sr.No	prev	prevgrade	attendence	internal	midterm	lab	sem	semgrade
1	74	O	good	good	67	72	73	O
2	72	O	avg	good	65	54	62	A
3	51	B	good	avg	61	62	58	B
4	62	A	avg	good	65	63	60	B
5	76	O	good	good	71	78	81	E
6	73	O	good	avg	65	67	70	A
7	78	O	good	poor	63	62	61	A
8	73	O	avg	good	68	60	68	A
9	74	O	avg	avg	65	58	62	A
10	71	O	avg	poor	62	72	62	A
11	52	B	poor	good	64	65	58	B
12	54	B	poor	avg	55	56	57	B
13	55	B	poor	poor	52	58	51	B
14	64	A	poor	good	62	62	61	A
15	69	A	poor	avg	58	61	59	B
16	61	A	poor	poor	53	57	58	B
17	75	O	poor	good	64	65	62	A
18	71	O	poor	avg	64	65	63	A
19	70	A	poor	poor	62	69	65	A
20	57	B	avg	good	57	61	61	A
21	51	B	avg	avg	57	62	60	B
22	52	B	avg	poor	54	52	43	X
23	57	B	good	good	61	68	63	A
24	53	B	good	avg	54	61	57	B
25	55	B	good	poor	56	58	54	B
26	64	A	avg	good	65	68	63	A
27	62	A	avg	avg	62	69	61	A
28	65	A	avg	poor	67	62	59	B
29	64	A	good	good	68	84	71	O
30	67	A	good	avg	64	63	66	A
31	68	A	good	poor	63	61	59	B

Figure 3.2   Sample dataset.

3.5.2 Data Preprocessing and Transformation

Following are the list of issues in data collection:

- Real world data cannot be collected perfectly, errors will occur always
- Interpretation of data is the issue faced by data miner
- Accessibility of right data for analysis often an issue
- Accessing the most recent data can sometimes a challenge

Preprocessing includes finding incorrect or missing data. It is applied for identifying the misplaced values, noisy data, and unrelated and unneeded information from dataset. Missing data must be supplied whereas erroneous data may be corrected or removed. The conversion of data into a common format for processing is known as Transformation. Some data may be encoded into more usable format. Thus, data reduction, dimensionality reduction, and data transformation method can be used to reduce the number of possible data values being considered.

While building data, it is required to improve data. It includes cleaning up the data, handling missing values, transforming data, and analyzing data to increase its efficiency through a proper choice of variables, that is, we need to transform raw data to the polished form to build data mining models.

As part of the data preprocessing and to get better input data for data mining some preprocessing tasks are applied for the collected data prior to loading the dataset to the data mining tool, unwanted attributes should be removed. There is a possibility of error in collected data. Simple data entry errors occur frequently. Data cleaning has to be done before building models. Proper preprocessing techniques must be applied on dataset before working with the original data such as:

1. Missing data is common feature of any dataset. Missing values replaced with sentinels to mark that they are missing. The omitted data were substituted using mean imputation method, that is, replacing the missing value for a given field by the mean of all known values of that field in the class.

2. Sampling: A sub-sample from the original dataset was taken for the use of bringing balance into training and testing sets. Shuffled sampling technique is applied on the subsample where a chosen item cannot be chosen again.

3. Feature selection deals with the selection of only the attributes which could be relevant for prediction. Generally, not all the attributes in a database will be used while mining. A group of forward selection and backward elimination was used to select the best features and to remove worst features. Selection of variables is important in improved modeling. It includes by selecting different subsets of variables and exploring them for best results. Techniques used for this purpose are as follows: decision tree induction, random forest, principle component analysis, etc.

4. Outlier is an observation that has quite different values from most other observations, can have an adverse impact on quality of result. Such outliers are detected using outlier detection algorithms and one approach is to handle it is removing them from the dataset.

5. Set Role operator in rapid miner indicates the role played by each attribute while classification. The variable to be predicted was given a role of label, and other variables were treated as regular.

Data pre-processing applied for recognizing the missing values, noisy data, and unrelated and redundant information from dataset. There are some R mining algorithms works on numeric values only, categorical variables are converted in to numerical form. Categorical variables such as "attendance", "internals", and "grade" are converted to numerical form as explained in the Tables 3.2 and 3.3.

3.5.3 R Packages and Functions for Data Mining

After preprocessing the data, research applies data mining techniques—association, classification, clustering, etc. for the analysis of educational data. In each of these tasks, research will present the extracted knowledge and describes its importance in educational domain. Table 3.4 lists the collection of R packages and function used for data mining. These data mining techniques and methods need brief mention for understanding:

- Classification makes use of a set of pre-classified examples to build a model that can classify the set of records at large. This technique employs decision tree or neural network-based classification algorithms. Classification models are designed in this research with R packages *rpart* and *nnet*. Various classifiers such as decision tree, neural network, Naïve Bayes are designed to classify the data. Classifier rules will be designed by applying classification algorithms on testing data.
- Clustering is an identification of similar classes of objects. It identifies dense and sparse regions in data space and can find out overall distribution pattern and correlations between data attributes. Various clustering

Table 3.2   Categorical variables "attendance" and "internals" numerical conversion

Attendance/Internals	Numerical Assignment
Good	1
Average	2
Poor	3

Table 3.3   Dépendent variable "Grade" description

Description (Grade)	Range	Preprocessed Value
Super 'S'	[91–100]	1
Exemplary 'E'	[81–90]	2
Outstanding 'O'	[71–80]	3
Good 'A'	[61–70]	4
Average 'B'	[51–60]	5
Defaulter 'X'	Less than 50	6

Table 3.4   Collection of R packages and function used for data mining [24]

Type	Package	Function Details
Clustering	*fpc*	Partitioning-based clustering:
	cluster	*kmeans, pam, pamk, clara*
	pvclust	Hierarchical clustering: *hclust,*
	mclust	*pvclust, agnes, diana*
		Model-based clustering: *mclust*
		Density-based clustering: *dbscan*
		Plotting cluster solutions:
		plotcluster, plot.hclust
		Validating cluster solutions:
		cluster.stats
Classification	*rpart*	Decision trees: *rpart, ctree*
	party	Random forest: *cforest,*
	randomForest	*randomForest*
	rpartOrdinal	Regression, Logistic regression,
	tree	Poisson regression: *glm, predict,*
	marginTree	*residuals*
	maptree	Survival analysis: *survfit, survdiff,*
	survival	*coxph*
Association Rules and	*arules*	APRIORI Algorithm: *apriori, drm*
Frequent Itemsets	*drm*	ECLAT Algorithm: *éclat*
Sequential Patterns	*arulesSequences*	SPADE algorithm: *cSPADE*
Time Series	*timsac*	Time series construction: *ts*
		Decomposition: *decomp,*
		decompose, stl, tsr
Statistics	*nlme*	Analysis of Variance: *aov, anova*
		Density analysis: *density*
		Statistical test: *t.test, prop.test,*
		anova, aov
		Linear mixed-e ects model t: *lme*
		Principal components and factor
		analysis: *princomp*
Graphics		Bar chart: *barplot*
		Pie chart: *pie*
		Scattered plot: *dotchart*
		Histogram: *hist*
		Density: *densityplot*
		Candlestick chart, box plot: *boxplot*
		Tree: *rpart*
Data Manipulation		Missing values: *na.omit*
		Standardize variables: *scale*
		Sampling: *sample*
		Others: *aggregate, merge, reshape*

techniques including K-means clustering, k-medoids clustering, and density-based clustering are applied. Comparative study of algorithms for various models will carried out to select the efficient one.

- Association and correlation is generally to find frequent item set findings within large datasets. This type of finding helps to make certain decisions. Association rule mining will be applied using *arules* package to find dependency factors.

3.5.4 Result Evaluation and Knowledge Presentation

The discovered knowledge is used for prediction of students' performance and constructive recommendation to conquer the problem of low grades of students. The model development, building classifiers, and their evaluation are explained in Chapter 4. Educational data analysis using clustering has explained in Chapter 5. Further result evaluation, knowledge presentation and results of performance prediction covered in Chapter 6. This framework can be used as a base for the prediction of students' performance in campus placement.

3.6 Working with Data

Data are the starting point for all data mining. Collection of data named as dataset includes set of observations. These observations are also named as entities, records, etc. Columns are named as variables, attributes, fields, etc. Variables or attributes can serve as different roles:

1. Input or dependent variables
2. Output or target or independent variables

Depending on the type of the value stored in the columns, there are two types of variables:

1. A categorical variable is one that takes on a single value, for a particular observation, from a fixed set of possible values

   Ex: color

2. A numeric variable has values that are integers or real numbers

   Ex: bank balance

While building classifiers, a dataset is partitioned into three independent datasets. This partitioning is done randomly so that each dataset represents the whole collection of observations:

1. The model is built by **training** dataset
2. Model's performance is validated by **validation** dataset
3. Model's performance assessed by **testing** dataset

In R, dataset named as data frame. A data frame is a list of variables where each variable in the list represents a column of data values. R refers a variable within a dataset as a vector. After collecting data, the next task is to structure the data into a form appropriate for data mining. In this case, this involves putting the data into a form that allows it to be loaded into R, Rattle. We will then explore, test, and transform this dataset in various ways for mining.

3.7 Research Methodology

Data Mining is the task being carried out, to generate the desired result. This research on mining of educational data deals with various algorithms in order to find out and extract patterns of stored data. Present research attempted to design data mining model in R, a free software environment. R is a simple, but very powerful data mining and statistical data processing tool for research. The mining program presented here was implemented and tested using R version 3.1.2, and the scripts are accessible to the reader as additional materials. The R packages and functions used in the data mining process and the corresponding additional materials are summarized in Table 3.5.

The main objective of the proposed methodology is to build the classification model that tests certain attributes that affect students' performance. To accomplish this, the DM methodology is used to build a classification model. Data pre-processing applied for identifying the misplaced values, noisy data, and unrelated and redundant information from dataset. As classification algorithms works on numeric values only, categorical variables are converted in to numerical form.

"Grade", semester examination performance is classified into six classes as explained in Table 3.3. For the purpose of supervised, learning the class variable "Grade" is treated as dependent variable. Table 3.6 gives students' grade and corresponding number of observations in dataset considered in this research.

Table 3.5   Methodology details of EDM in R

Step	Aim	Method	R Package	Supplementary Material
1	Data Collection	Student details from examination section and department of Computer Studies.	—	student.xls
2	Data Preprocessing	Categorical values converted in to numerical	—	student.Rdata
3	Data Extraction and Data Exploration	Descriptive Statistics	Hmisc, ggplot	exploration.r
4	Building Classifiers	Decision Tree Random Forest Neural Network K-Nearest Neighbor Naïve Bayes	Party, rpart randomForest Nnet Class, e1071	classifiers.r
5	Creating Clusters	K-means Hierarchical Hybrid Hierarchical Density Based	Cluster hybridHclust Stats Fpc, E107	Clustering.r
6	Model Selection	Confusion Matrix	nnet	selection.r

Table 3.6   Students' grade and number of observations in dataset

Description (Grade)	No. of Observations
Super 'S'	0
Exemplary 'E'	11
Outstanding 'O'	25
Good 'A'	81
Average 'B'	63
Defaulter 'X'	20

3.8 Loading and Exploring Data—Exploratory Data Analysis

Data can appear in different formats from different sources. R supports importing data in many formats. In R data can be imported from Excel, SAS, SPSS, Oracle, etc. Using R's extensive capabilities, Rattle provides direct access to such data. Dataset *EDM.csv* is loaded by giving following command. It creates a data frame "*student*".

student<- read.csv("D:/Rmining/EDM.csv")

Rattle can load dataset from various sources, such as CSV (comma separated data), TXT (tab separated data), ARFF (a common data mining dataset format), and ODBC connections R data frames attached with current R session, and the packages installed in the R libraries, are also available through the Rattle interface. The **spreadsheet** option of the **Data** tab provides the simplest way of loading data into Rattle. Corresponding loading of *EDM.csv* in Rattle is shown in Figure 3.3. It shows the list of available variables and their default roles.

Even before building data mining models, significant insights can be gained by exploring the dataset. Exploratory data analysis is a core activity in data mining projects. It involves basic understanding of dataset through summaries and visual plots. Table 3.7 lists set of R commands for exploring data and corresponding interpretation.

The purpose of exploratory analysis of data is to highlight constructive information and support decision making. In the educational organization, for example, it can assist educators to analyze the students' activities and to get a

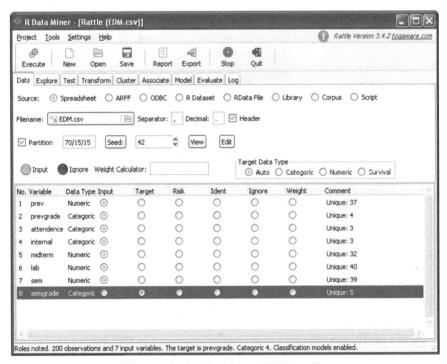

Figure 3.3   Loading dataset EDM.csv file.

Table 3.7 Exploratory data analysis in R

Commands	Interpretation with Example
student<- read.csv ("D:/Rmining/EDM.csv")	Creates a dataframe *"student"* by loading dataset from *EDM.csv*.
print(student)	Displays dataset in R console.
dim(student)	Returns dimensions of dataset Eg. 200 observations and 9 variables
names(student)	Displays variable names or column names Eg: "Sr.No" "prev" "prevgrade" "attendence" "internal" "midterm" "lab" "sem" "semgrade"

str(student)	Displays structure of *student* dataframe
	'data.frame': 200 obs. of 9 variables:
	$ Sr.No: int 1 2 3 4 5 6 7 8 9 10 ...
	$ prev: int 74 72 51 62 76 73 78 73 74 71 ...
	$ prevgrade: Factor w/ 4 levels "A","B","E","O":
	4 4 2 1 4 4 4 4 4 4 ...
	$ attendence: Factor w/ 3 levels
	"avg","good","poor": 2 1 2 1 2 2 2 1 1 1 ...
	$ internal: Factor w/ 3 levels
	"avg","good","poor": 2 2 1 2 2 1 3 2 1 3 ...
	$ midterm: int 67 65 61 65 71 65 63 68 65 62 ...
	$ lab: int 72 54 62 63 78 67 62 60 58 72 ...
	$ sem: int 73 62 58 60 81 70 61 68 62 62 ...
	$ semgrade: Factor w/ 5 levels
	"A","B","E","O",..: 4 1 2 2 3 1 1 1 1 1 ...

attributes(student)	Displays variable names and other details $names "Sr.No" "prev" "prevgrade" "attendence" "internal" "midterm" "lab" "sem" "semgrade"
summary(student)	Gives distribution and summary of every variable as shown in Figure 3.5. It shows basic text based statistical summary of the dataset. For numeric variables it lists minimum, maximum with average values. For categorical variables it lists frequency counts.
table(student$semgrade)	Gives frequency of class variable *"semgrade"* A B E O X 81 63 11 25 20

general view of a student's learning. The two main techniques mostly used for this task are statistics and visualization. R codes and corresponding outputs explained in this section.

Statistics is a mathematical science focusing on the collection, study, interpretation, and presentation of data. It is relatively simple to get basic descriptive statistics from R. This descriptive analysis can provide summaries and reports about learner behavior when it is used with educational data. Data exploration with R starts with inspecting the dimensionality, structure, and data of an R object. Script written in R for Exploration and Visualization of dataset is shown in Figure 3.4. Output of this is shown in Figure 3.5. Dataset contains six classes of students based on semester end examination performance. There are 200 observation and 6 columns, *"Grade"* is target variable. Data frame is created by reading dataset. Results of basic statistical computation and number of observations for each type are reported by summary function as given Figure 3.6.

3.9 Interactive Graphics and Data Visualization

Graphical tools allow visually exploring the dataset's characteristics to help us in understanding it. Distribution of values can be reviewed visually during the beginning of data mining project. R is one of the data visualization languages, has many options for graphically presenting data. Raw data and algorithm results can be visualized through graphics such as histograms and density diagrams.

The aim was to present data along certain parameters and make extreme points and clusters noticeable to human eye. Information visualization uses graphic tools to help people understand and analyze data. Grade density of different categories is shown in Figure 3.7a–c and corresponding R code for the same is given. To plot frequency of class variable *"semgrade"*, values must

```
# data extraction and exploration(student)
student<- read.csv("G:/R_systems/Rmining/EDM.csv")
attributes(student)
str(student)
summary(student)
plot(student$semgrade)
plot(density(student$semgrade))
table(student$semgrade)
pie(table(student$semgrade_num))
```

Figure 3.4   R script for exploration and visualization.

```
R Console                                                            _|□|×|
> student<- read.csv("G:/R_systems/Rmining/EDM.csv")
> attributes(student)
$names
[1] "Sr.No"      "prev"       "prevgrade" "attendence" "internal"
[6] "midterm"    "lab"        "sem"       "semgrade"

$class
[1] "data.frame"

$row.names
  [1]    1   2   3   4   5   6   7   8   9  10  11  12  13  14  15  16  17  18
 [19]   19  20  21  22  23  24  25  26  27  28  29  30  31  32  33  34  35  36
 [37]   37  38  39  40  41  42  43  44  45  46  47  48  49  50  51  52  53  54
 [55]   55  56  57  58  59  60  61  62  63  64  65  66  67  68  69  70  71  72
 [73]   73  74  75  76  77  78  79  80  81  82  83  84  85  86  87  88  89  90
 [91]   91  92  93  94  95  96  97  98  99 100 101 102 103 104 105 106 107 108
[109]  109 110 111 112 113 114 115 116 117 118 119 120 121 122 123 124 125 126
[127]  127 128 129 130 131 132 133 134 135 136 137 138 139 140 141 142 143 144
[145]  145 146 147 148 149 150 151 152 153 154 155 156 157 158 159 160 161 162
[163]  163 164 165 166 167 168 169 170 171 172 173 174 175 176 177 178 179 180
[181]  181 182 183 184 185 186 187 188 189 190 191 192 193 194 195 196 197 198
[199]  199 200

> str(student)
'data.frame':   200 obs. of  9 variables:
 $ Sr.No     : int  1 2 3 4 5 6 7 8 9 10 ...
 $ prev      : int  74 72 51 62 76 73 78 73 74 71 ...
 $ prevgrade : Factor w/ 4 levels "A","B","E","O": 4 4 2 1 4 4 4 4 4 4 ...
 $ attendence: Factor w/ 3 levels "avg","good","poor": 2 1 2 1 2 2 2 1 1 1 ...
 $ internal  : Factor w/ 3 levels "avg","good","poor": 2 2 1 2 2 1 3 2 1 3 ...
 $ midterm   : int  67 65 61 65 71 65 63 68 65 62 ...
 $ lab       : int  72 54 62 63 78 67 62 60 58 72 ...
 $ sem       : int  73 62 58 60 81 70 61 68 62 62 ...
 $ semgrade  : Factor w/ 5 levels "A","B","E","O",..: 4 1 2 2 3 1 1 1 1 1 ...
```

Figure 3.5   Output of R code for exploration.

```
R Console                                                            _|□|×|
> summary(student)
     Sr.No             prev        prevgrade attendence internal    midterm
 Min.   :  1.00   Min.   :50.00   A:66      avg :70    avg :67   Min.   :50.0
 1st Qu.: 50.75   1st Qu.:57.00   B:67      good:72    good:79   1st Qu.:58.0
 Median :100.50   Median :65.00   E: 9      poor:58    poor:54   Median :63.0
 Mean   :100.50   Mean   :64.96   O:58                           Mean   :63.2
 3rd Qu.:150.25   3rd Qu.:72.00                                  3rd Qu.:67.0
 Max.   :200.00   Max.   :87.00                                  Max.   :86.0
      lab             sem          semgrade
 Min.   :43.0    Min.   :43.00    A:81
 1st Qu.:58.0    1st Qu.:58.00    B:63
 Median :62.5    Median :62.00    E:11
 Mean   :63.3    Mean   :62.73    O:25
 3rd Qu.:68.0    3rd Qu.:68.00    X:20
 Max.   :86.0    Max.   :86.00
>
```

Figure 3.6   Basic statistical computation details.

plot(hist(student$grade_num))

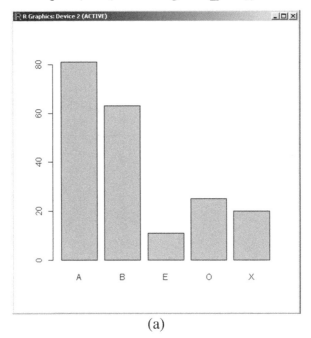

(a)

plot(density(student$grade_num))

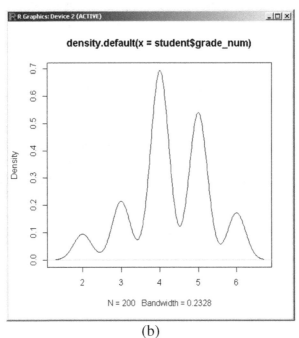

(b)

pie(table(student$grade_num))

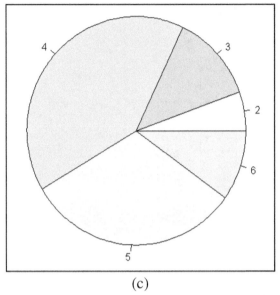

(c)

Figure 3.7   (a) Histogram for class variable "*semGrade*". (b) Density plot for class variable "*semGrade*". (c) Pie chart for class variable "*semGrade*".

be in numeric form. Corresponding numeric values given in Table 3.3. Results can be visualized through particular techniques such as symbolic data analysis where groups are created by gathering individuals along one attribute.

3.10 Conclusion

This chapter covered research design and methodology adopted to carry out EDM study and analysis. A step wise procedure for research framework design is defined and executed. The study revealed and identified the key factors that determine the outcome of students' performance. Corresponding dataset collected form CSIBER, a postgraduate institute. Data pre-processing applied as per the requirement of R mining algorithms. Functions and packages in R are selected and installed for the purpose of mining. Statistics and visualization information retrieved from student dataset are provided. Various EDM methods and algorithms are adopted in this research to discover hidden patterns and relationships.

4

Model Development—Building Classifiers

4.1 Introduction—Descriptive and Predictive Analytics

Modeling is the process of considering some data and building the simplified description of the processes that have produced it. Model can be an algorithm or a mathematical formula. It captures knowledge exhibited the data and presents it in some form. Algorithm for building a model is referred as model builder. R and Rattle supports various model builders such as clustering, classification, association. Building model focuses on learning relationships between the input variables and target variable.

Data mining model building algorithms are categorized in to descriptive analytics and predictive analytics. In a traditional machine-learning context, these equate to unsupervised learning and supervised learning. In descriptive analytics, the discovered knowledge is represented without necessarily modeling a specific outcome. The tasks of cluster analysis, association and correlation analysis, and pattern discovery are fall under this category. In predictive analytics, the models will extract knowledge from historic data and the same can be applied to new situations. From a machine-learning perspective, this is also referred to as supervised learning. The task of classification and regression are fall under this category. Classifiers are used to predict the category of new observations. Regression models used to predict a numeric outcome. This chapter describes predictive analytics algorithms and its usage in R/Rattle with illustrative examples. The objective was to provide insight into how these techniques work so that as a data miner can make effective use of the models.

4.2 Predictive Analytics

Prediction of a student's performance is one of the oldest and most popular applications of data mining in education, and different techniques and models have been applied. The objective of prediction is to estimate the unknown value of a variable that describes the student. In education, the values normally predicted are performance, knowledge, score or mark. Classification is a supervised learning technique that classifies record into predefined class label. This is one of the most constructive techniques in data mining to build classifiers from an input dataset. With classification, the designed model will be able to forecast a class for given data item depending on earlier learned information.

Classification and regression are two types of prediction. This value can be numerical/continuous value or categorical/discrete value. Regression analysis finds the relationship between a dependent variable and one or more independent variables. In regression analysis continuous input variables are used. Classification is a technique in which individual items are located into groups based on quantitative information regarding one or more characteristics inherent in the items and based on a learning set of previously labeled items. In case of classification, the categorical or binary variables are used, but in regression continuous input variables are used. In this research, data mining algorithms were utilized to build classifiers to predict the performance of students.

In general, the classification of dataset is a two-step process. In the former step, which is called the learning step, a model that describes a predetermined set of classes or concepts is built by analyzing a set of training database instances. Each instance is understood to belong to a predefined class. In the second step, the model is tested using a different dataset that is used to estimate the classification accuracy of the model. If the accuracy of the model is considered good enough, the model can be used to categorize future data instances for which the category label is not known. At the end, the model acts as a classifier in the decision-making process. This chapter explains different data mining techniques and algorithms for classifying students details based on their academic performance in semester end examination.

4.3 Dataset and Class Labels

This research intent in extraction of the hidden knowledge from the student database and perform prediction of students' academic performance. Based on the study, five parameters were selected to determine the grade, that is, students' performance in semester end examination. These selected parameters include.

- Previous semester result
- Performance in class test

- Grade scored in assignment
- Attendance
- Performance in practical examination

This research has collected real-time data that describing the relationships between learning behavior of students and their academic performance. Sample data are illustrated in Figure 3.2 in Chapter 3. The dataset contain students' details which have been recorded and subjected to the data mining process. Students' academic performance in semester end examination is separated into six classes as:

1. Super "S"
2. Exemplary "E"
3. Outstanding "O"
4. Good "A"
5. Average "B"
6. Defaulter "X"

In the case educational domain, it is also very significant for the classifier obtained to be user friendly, so that educator can make decisions about students to improve the students' learning.

4.4 Classification Framework and Process

The present research carried out a data mining techniques implementation using different R packages. Classification process framework is shown in Figure 4.1. A three-step procedure is followed for the design of research framework:

1. Develop different classifiers that can be used to predict performance outcome. It consists of a supervised learning phase where the classifier models the association between the target variable "grade" and the explanatory variables using a training set randomly selected from the 200×6 dataset.
2. Evaluate the performance of the each classifier and select the "best" one. It allows checking the performance of the trained classifiers against a testing set, evaluating their predictive accuracy with data not used in the training step.
3. Selected classifier is deployed for the early prediction of students' performance.

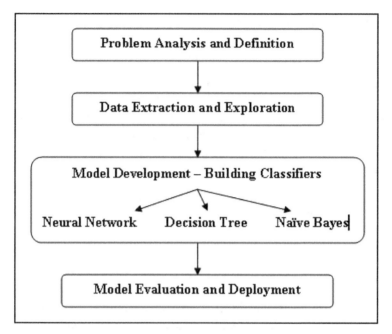

Figure 4.1   Classification framework.

The classification algorithms commonly build models which can be used to predict future data drifts. There are several techniques for data classification such as Naïve Bayes classifiers, neural network, and decision tree.

Classification is a data mining task that maps the data into predefined groups & classes. It is also called as supervised learning. It consists of three steps:

1. Model construction: It consists of set of predetermined classes. Each data item is assumed to fit in to a predefined class. The set of sample used for model construction is learning or training set. The model is represented as neural network, decision trees, or mathematical principles. This model is shown in Figure 4.2.

2. Model usage: This model is used for classifying future or unknown objects. The known label of test data item is compared with the classified result of the classifier. Accuracy rate is the percentage of test set that are properly classified by the classifier. Test set is independent of learning set, otherwise over-fitting may occur. This model is shown in Figure 4.3.

3. Prediction: It is used to model continuous-valued functions, that is, predicts unknown or missing values. In this model, single aspect of data deduced from some combination of other aspect of data. In educational data mining, prediction can be used to detect student behavior, predicting, or understanding student educational outcomes. This model is shown in Figure 4.4.

In educational data mining, given details of a student, one may predicate his/her final grade. The decision tree is used to symbolize logical rules of students' final grade.

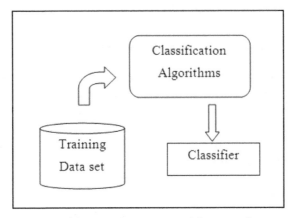

Figure 4.2   Learning step or model construction.

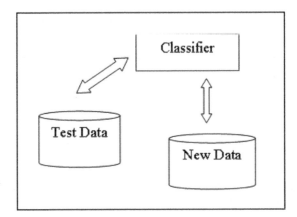

Figure 4.3   Model usage.

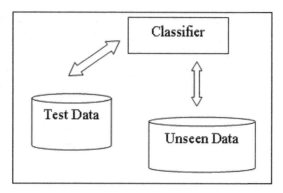

Figure 4.4   Prediction model.

4.5 Predicting Students' Performance

Predicting the performance of a student is a great concern to the higher education managements. In real world, predicting performance of a student is a challenging task. In EDM, classifiers are the most common types of prediction models. Classification is the most generally applied data mining method, which employs a group of pre-classified examples to design a model that can categorize the population of records at large. In classifiers, the predicted variable can be either a binary or categorical variable. Decision trees, random forest, decision rules, step regression, and logistic regression are some popular classification methods in educational domains include.

There have been numerous studies done in constructing predictive models for various purposes in the education field. For example, the study has been done by attempting to cluster students into three groups such as strong ability, medium ability, and low ability.

- The "strong ability", students with a high probability of succeeding until the last semester.
- The "medium's ability" and "low ability", students who may need help from educators.
- For "low ability" students, the educators can pay more attention to them by giving them extra lesson and assignment to improve their performances.

To produce an effective predictive model, it requires good input parameter, suitable data mining methods and tools for the data. Prediction methods in EDM have two main objectives; to study what parameters of a model are important for predicting, and to give information about underlying construct.

4.6 Classification and Predictive Modeling in R and Rattle

Classification is the most generally applied data mining method, which uses a set of pre-classified examples to design classifier. This approach commonly employs neural network or decision tree-based classification algorithms. The classification process includes learning and classification. During learning phase, the training data are analyzed by classifier. In classification step, test data are used to approximate the accuracy of the classifier. If the accuracy is satisfactory, the model can be applied to the new data items. The classifier-learning algorithm uses pre-classified examples to find out the set of attributes required for appropriate discrimination. The algorithm after that encodes these attributes into a model called a classifier.

The goal of classification method was to classify the students' dataset to the predefined classes or groups. The method is a supervised technique and it is also called direct learning because the classes are predefined before extracting patterns on the target data. Classification techniques such as Decision Tree, NBC, and Neural Network are the most frequently applied techniques in EDM.

Classification is a supervised learning technique that classifies data into predefined class label. This is the most useful method in data mining to build classifiers from an input dataset, which can be used to predict future data trends. Various classification methods implemented in this research are explained in Section 4.7–4.10.

Data mining model is developed by building classification rules for the categorical target variable "*grade*". Three classification methods during the learning step are chosen to represent a wide range of approaches in statistics. Analysis starts with a random split of the dataset into two parts: a learning set and a testing set. The learning set *studTrain* contains 140 observations used to train the three classifiers. The remaining 60 observations *studVal* are used to evaluate the "out-of-sample" performance of the classifiers. This has done by sample command as given.

$$studTrain = sample(1:200, 140)$$
$$studVal = setdiff(1:200, studTrain)$$

4.7 Decision Tree Modeling

Decision trees are traditional building blocks of data mining. Construction of a tree begins with a single root node that splits into multiple branches, leading to additional nodes, branching continuous or else terminate as a leaf node. Each non-leaf node associated with a test condition that decides which branch to follow. The leaf nodes contain the decisions.

Thus, a decision tree is a set of conditions arranged in a hierarchical structure. This is a predictive model in which a data item is categorized by following the path of fulfilled conditions from the root of the tree till reaching a leaf. The leaf corresponds to a class label. A set of classification rules can be easily derived from decision tree. It is a quite popular technique because the construction of decision tree does not need any domain expert knowledge and is suitable for exploratory knowledge discovery.

Decision tree is a very convenient and efficient way of representing knowledge. Model expressed in one language can be translated to another language. Simple and useful translation is using rule set. A rule representation is beneficial in reviewing the knowledge that has been captured. It is also easy to translate each rule into a programming language statement as an "if-then" statement.

In a decision tree, each branch node represents a choice among a number of alternatives, and each leaf node denotes a decision. It starts with a root node where users can take actions. From root node, users divide each node recursively according to learning algorithm. A decision tree is the one which each branch represents a possible scenario of decision.

4.7.1 Decision Tree Implementation in R

The *rpart* package is used for classification by decision trees and can also be used to generate regression trees. Recursive partitioning is a fundamental tool in data mining. It explores the structure of a set of data, while developing easy to visualize decision rules for predicting a categorical. The resulting models can be represented as binary trees. R implementation of decision tree is shown in Figure 4.5. Corresponding output in the form of confusion matrices is shown in Figure 4.6.

It is one of the most used techniques, since it creates the decision tree from the data given using simple equations depending mainly on calculation of the gain ratio, which gives automatically some sort of weights to attributes used, and the researcher can implicitly recognize the most effective attributes on the predicted target. As a result of this technique, a decision tree would be built with classification rules generated from it. Figure 4.8 shows decision tree generated for the inputted *student* dataset.

```
#Decision Tree
student<- read.csv("G:/R_systems/Rmining/EDM_classifier.csv")
student[stTrain,]
student[stVal,]
table(student[stTrain,]$grade)
table(student[stVal,]$grade)
library(party)
library(grid)
library(mvtnorm)
library(modeltools)
library(stats4)
library(strucchange)
library(zoo)
library(sandwich)
formula <- grade_num ~ prev+att_num+internal_num+midterm+lab
stud_ctree <- ctree(formula, student[stTrain,])
print(stud_ctree)
plot(stud_ctree)
text(stud_ctree, use.n=T)
table(predict(stud_ctree, newdata = student[stTrain,]),
student[stTrain,]$grade_num)
table(predict(stud_ctree, newdata = student[stVal,]), student[stVal,]$grade_num)
```

Figure 4.5 R implementation of decision tree.

The tree is built by the following process:

1. First the single variable is found which best splits the data into two groups
2. The data is separated, and then, this process is applied separately to each sub-group, and
3. So on recursively until the subgroups either reach a minimum size or until no improvement can be made.

It can produce a model with rules that are human-readable and interpretable. Rules generated in this implementation of decision tree are given in Figure 4.7. This model has the advantages of easy interpretation and understanding for decision makers to compare with their domain knowledge for validation and justify their decision. C4.5/C5.0/J4.8, CART, NBTree, etc are some of the well-known decision tree algorithms.

```
> table(student[stTrain,]$grade)

 2  3  4  5  6
10 17 55 44 14
> table(student[stVal,]$grade)

 2  3  4  5  6
 1  8 26 19  6

table(predict(stud_ctree, newdata = student[stTrain,]), student[stTrain,]$grade_num)

                    2  3  4  5  6
2.33333333333333   10  5  0  0  0
3.30769230769231    0  9  4  0  0
3.83333333333333    0  2 10  0  0
4.09090909090909    0  0 10  1  0
4.17647058823529    0  1 12  4  0
4.45454545454545    0  0  6  5  0
4.80434782608696    0  0 12 31  3
5.66666666666667    0  0  1  3 11
table(predict(stud_ctree, newdata = student[stVal,]), student[stVal,]$grade_num)

                    2  3  4  5  6
2.33333333333333    1  1  0  0  0
3.30769230769231    0  3  2  0  0
3.83333333333333    0  2  2  0  0
4.09090909090909    0  0  2  3  0
4.17647058823529    0  1 15  1  0
4.45454545454545    0  1  3  4  0
4.80434782608696    0  0  2 10  4
5.66666666666667    0  0  0  1  2
```

Figure 4.6   Output of decision tree implementation in R.

4.7.2 Decision Tree in Rattle

The *student* dataset is used to illustrate the building of decision tree in Rattle. The preprocessed nominal dataset is loaded into Rattle by selecting data tab. Corresponding execution is show in the Figure 4.9. Rattle randomly partition the whole dataset into train/validate/test datasets. Partition is created each time this tab is executed. If partition is enabled, modeling is performed over the training dataset. In the evaluation tab, the default dataset for evaluation will be in the validate dataset. The test dataset is used to obtain an unbiased error estimate.

After loading *student* dataset and identifying the input and the target variables, an *Execute* of the Model tab will result in a decision tree. The result of this execution in text view is shown in Figure 4.10. This textual summary includes much information. Clicking on the *Rules* button gives a set of rules at the bottom of the text view. Draw button of the interface will pop up a window, displaying the pictorial representation of decision tree as shown in Figure 4.11. Error can be reduced by re-portioning the dataset.

```
> formula <- grade_num ~ prev+att_num+internal_num+midterm+lab
> stud_ctree <- ctree(formula, student[stTrain,])
> print(stud_ctree)

          Conditional inference tree with 8 terminal nodes

Response:  grade_num
Inputs:    prev, att_num, internal_num, midterm, lab
Number of observations:   140

1) lab <= 63; criterion = 1, statistic = 96.199
  2) lab <= 52; criterion = 1, statistic = 28.917
    3)* weights = 15
  2) lab > 52
    4) midterm <= 62; criterion = 0.998, statistic = 12.36
      5)* weights = 46
    4) midterm > 62
      6) internal_num <= 1; criterion = 0.964, statistic = 7.202
        7)* weights = 11
      6) internal_num > 1
        8)* weights = 11
1) lab > 63
  9) lab <= 74; criterion = 1, statistic = 30.301
    10) att_num <= 1; criterion = 0.995, statistic = 10.684
      11) internal_num <= 1; criterion = 0.971, statistic = 7.578
        12)* weights = 13
      11) internal_num > 1
        13)* weights = 12
    10) att_num > 1
      14)* weights = 17
  9) lab > 74
    15)* weights = 15
```

Figure 4.7 Rules generated by decision tree.

Execution of *Evaluate* tab gives performance of decision tree model in-terms of error matrix as shown in Figure 4.12. Model can be evaluated for training, validation, and testing dataset.

4.8 Artificial Neural Network Classifier

Neural network is a set of connected input/output units and each connection has a weight present with it. During the learning phase, network learns by adjusting weights so as to be able to predict the correct class labels of the input tuples. Neural networks have the remarkable capability to derive meaning from imprecise data and can be utilized to extract patterns and identify trends that are complex to be detected by either humans or any computer techniques.

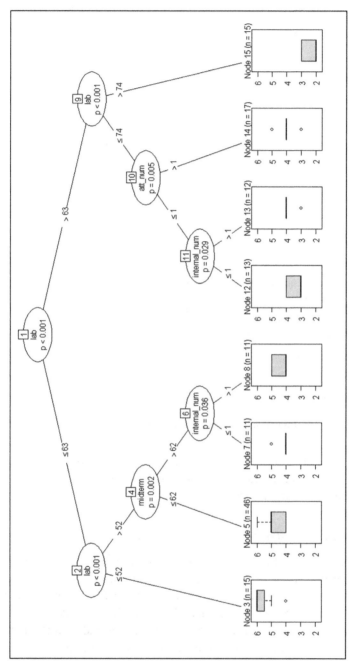

Figure 4.8 Decision tree for script given in Figure 4.5.

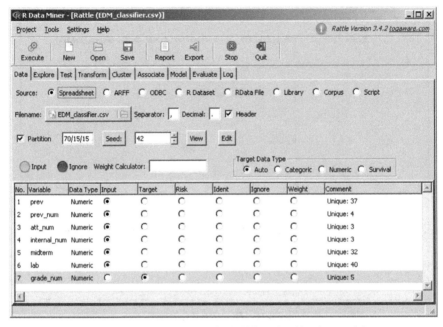

Figure 4.9 Loading dataset for building classification models.

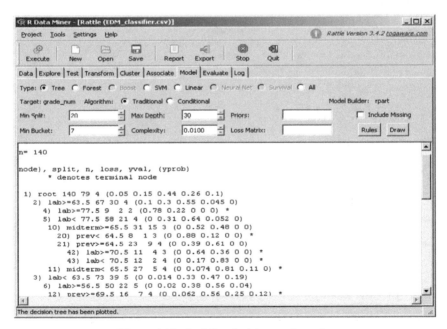

Figure 4.10 Building decision tree in rattle.

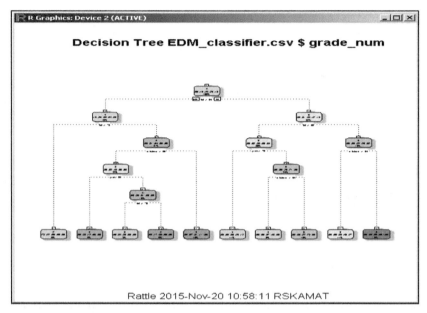

Figure 4.11   Decision tree in rattle.

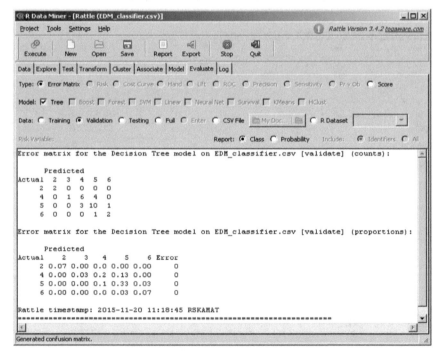

Figure 4.12   Decision tree performance.

These are well suited for continuous valued inputs and outputs. Neural networks are best at recognizing patterns or trends in data and well suited for prediction or forecasting needs.

Artificial neural network (ANN) is based on the human brain architecture that consists of multiple processing layers connected with nodes. ANN method is used in the educational field for predicting and classifying students based on their academic performance. Outstanding ability to derive meaning from complex data, extract patterns, and detect data trends are among the capability of ANN. These operations are sometimes too arduous to be carried out by human or ordinary calculation.

The package NNET provides methods for using feed-forward neural networks with a single hidden layer. For the purpose of machine learning, dataset is separated into training and validation sets. This allows validating the ANN on data that it was never trained with. The neural network requires that the records be normalized using one-of-n normalization. Input values are normalized between 0 and 1 as per the requirement of ANN. Training data is trained by *nnet* function. Generated model is tested on test data with predict function. R implementation of neural network is given in Figure 4.13. Output of *nnet* command and confusion matrices for ANN implementation in R is explained in Figure 4.14.

```
#Artificial Neural Network
student<- read.csv("G:/R_systems/Rmining/EDM_ann.csv")
table(student$grade_num)
stTrain = sample(1:200,140)
stVal = setdiff(1:200,stTrain)
student[stTrain,]
table(student[stTrain,]$grade_num)
student[stVal,]
table(student[stVal,]$grade_num)
library(nnet)
formula <- grade_num ~ prev+att_num+internal_num+midterm+lab
studANN = nnet(formula, data=student, subset=stTrain, size = 10, rang = 0.2,decay = 5e-4, maxit
              = 200)
table(student$grade_num[stTrain], predict(studANN, student[stTrain,], type = "class"))
table(student$grade_num[stVal], predict(studANN, student[stVal,], type = "class"))
predict(studANN, student[stTrain,], type="class")
predict(studANN, student[stVal,], type="class")
```

Figure 4.13 R implementation of neural network.

```
> table(student[stTrain,]$grade)

 A  B  E  O  X
55 44 10 17 14
> table(student[stVal,]$grade)

 A  B  E  O  X
26 19  1  8  6
> library(nnet)
> formula <- grade ~ prev+att_num+internal_num+midterm+lab
> studANN = nnet(formula, data=student, subset=stTrain, size = 10, rang = 0.2,decay = 5e-4, maxit = 200)
# weights:  115
initial  value 231.310135
iter  10 value 189.729213
iter  20 value 159.128014
iter  30 value 139.572610
iter  40 value 118.983075
iter  50 value 93.307456
iter  60 value 84.182273
iter  70 value 77.577468
iter  80 value 74.677791
iter  90 value 73.142096
iter 100 value 72.714869
iter 110 value 72.003702
iter 120 value 71.961285
iter 130 value 71.698763
iter 140 value 67.943963
iter 150 value 66.580939
iter 160 value 66.025725
iter 170 value 63.681508
iter 180 value 59.564721
iter 190 value 57.522732
iter 200 value 56.565583
final  value 56.565583
stopped after 200 iterations
```

```
> table(student$grade[stTrain], predict(studANN, student[stTrain,], type = "class"))

   A  B  E  O  X
A 46  7  0  2  0
B  8 36  0  0  0
E  0  0 10  0  0
O  2  0  0 15  0
X  1  2  0  0 11
> table(student$grade[stVal], predict(studANN, student[stVal,], type = "class"))

   A  B  E  O  X
A 14  5  0  3  4
B  6 11  0  0  2
E  0  0  1  0  0
O  3  0  0  5  0
X  1  4  0  0  1
> predict(studANN, student[stTrain,], type="class")
  [1] "B" "A" "A" "B" "A" "A" "B" "A" "B" "B" "A" "B" "B" "E" "B" "B" "A" "X"
 [19] "A" "B" "O" "A" "B" "A" "A" "A" "A" "B" "A" "B" "B" "A" "A" "B" "X" "A"
 [37] "O" "O" "A" "O" "A" "E" "X" "A" "A" "B" "E" "A" "E" "B" "A" "X" "E" "B"
 [55] "O" "B" "A" "A" "B" "B" "X" "E" "B" "A" "B" "A" "O" "X" "A" "O" "A" "O"
 [73] "A" "A" "A" "X" "A" "A" "O" "B" "X" "X" "B" "A" "O" "A" "B" "O" "B" "E"
 [91] "B" "B" "B" "O" "A" "B" "X" "A" "B" "A" "A" "E" "A" "A" "A" "B" "A" "B"
[109] "A" "E" "B" "B" "A" "A" "A" "A" "O" "B" "A" "B" "A" "O" "A" "O" "E" "B"
[127] "B" "O" "B" "A" "B" "A" "B" "A" "B" "A" "B" "A" "X" "O"
> predict(studANN, student[stVal,], type="class")
 [1] "A" "X" "A" "B" "A" "A" "B" "A" "A" "B" "B" "A" "A" "X" "A" "A" "B" "A" "B"
[20] "B" "A" "A" "A" "E" "X" "O" "B" "A" "O" "A" "A" "B" "B" "X" "B" "B" "B" "A"
[39] "X" "O" "O" "A" "A" "B" "B" "B" "O" "A" "O" "X" "B" "B" "A" "B" "B" "O" "O"
[58] "A" "X" "A"
```

Figure 4.14   ANN classification results in R.

4.9 Naive Bayes Classifier

A Naive Bayes classification technique is a probabilistic classifier based on applying Bayes' theorem with strong assumptions. Naïve Bayes classifier is another classification technique that is used to predict a target class. It depends in its calculations on probabilities, namely Bayesian theorem. Because of this use, results from this classifier are more accurate and effective, and more sensitive to new data added to the dataset.

NBC uses the Bayes' probability theory which assumes that the outcome of an attribute value of a given category is not dependent on the values of the other attributes. It represents a descriptive and predictive approach to predict the class membership for a target tuple. NBC is easy to use, capable to predict the unseen data, work well in various mediums with less error rate, and gives the highest accuracy compared to other methods. When applied to the existing students' datasets, predictive models can be generated to assist in the management of students' dropout, and to predict the performance of the new intake students.

The Naïve Bayes classification algorithm is based on a probabilistic model that incorporates set of strong independence assumptions. It can handle an arbitrary number of independent variables whether continuous or categorical. E1071 is a CRAN package is installed to execute Naïve Bayes Classifier. Figure 4.15 explains R implementation of Naïve Bayes classifier. Naïve Bayes Classification Results with confusion matrices shown in Figure 4.16. It gives expected training and validation set observations and corresponding confusion matrices.

4.10 Random Forest Modeling

A random forest is a collection of unpruned decision trees. It is often used when there is a very large training datasets and a very large number of input variables. This model is typically made up of tens or hundreds of decision trees. These models are generally competitive with nonlinear classifiers. However, performance is dependent on dataset taken for the study.

In random forest algorithm, each individual tree will over fit the data. Here, the randomness is in the selection of both observations and variables. This algorithm will often build from 100 to 500 trees. In case of deploying the model, the decisions made by each of the trees are combined by treating all trees as equals.

```
#naiveBayes classifier
library(class)
library(e1071)
student<- read.csv("G:/R_systems/Rmining/EDM_classifier.csv")
table(student$grade_num)
stTrain = sample(1:200,140)
stVal = setdiff(1:200,stTrain)
student[stTrain,]
table(student[stTrain,]$grade_num)
student[stVal,]
table(student[stVal,]$grade_num)
student$grade_num <- factor(student$grade_num)
table(student$grade_num)
classifier<-naiveBayes(student[stTrain,][,1:7], student[stTrain,][,8])
table(predict(classifier, student[stTrain,][,-8]), student[stTrain,][,8])
classifier<-naiveBayes(student[stVal,][,1:7], student[stVal,][,8])
table(predict(classifier, student[stVal,][,-8]), student[stVal,][,8])
```

Figure 4.15   R implementation of Naïve Bayes classifier.

```
table(student[stTrain,]$grade_num)

 2   3   4   5   6
 9  21  55  38  17

table(student[stVal,]$grade_num)

 2   3   4   5   6
 2   4  26  25   3

table(predict(classifier, student[stTrain,][,-8]), student[stTrain,][,8])

     2   3   4   5   6
 2   9   0   0   0   0
 3   0  21   3   0   0
 4   0   0  45   3   0
 5   0   0   7  34   1
 6   0   0   0   1  16

table(predict(classifier, student[stVal,][,-8]), student[stVal,][,8])

     2   3   4   5   6
 2   2   0   0   0   0
 3   0   4   0   0   0
 4   0   0  23   2   0
 5   0   0   3  23   0
 6   0   0   0   0   3
```

Figure 4.16   Naïve Bayes classification results in R.

4.10.1 Random Forest Model in R

The RANDOMFOREST package is used for classification by random forest classifiers. For classification, the corresponding method implements Breiman's random-forest algorithm. It can also be employed for assessing proximities among data points in unsupervised mode. Figure 4.17 explains R implementation of Random Forest classifier. Output of *randomforest* command and confusion matrices for implementation in R is explained in Figure 4.18.

4.10.2 Random Forest Implementation in Rattle

The dataset is loaded in Rattle for building random forest model as shown in Figure 4.9. After loading student dataset and identifying the input and the target variables, select *forest* button as type of a model. *Execute* of the Model tab will result in a random forest in text view. The result of this execution in text view is shown in Figure 4.19. This textual summary includes much information. Clicking on the Rules button gives a set of rules at the bottom of the text view as shown in Figure 4.20. A useful tool is error plot, obtained by clicking *Errors* button. Figure 4.21 shows resulting error plot for random forest model. The graph reports accuracy of the forest of trees against the number of trees included in the forest.

4.11 Model Selection and Deployment

After developing classifiers, the criteria defined for evaluating their performances. In this section, let us consider the issue of evaluating the performance of the models that have been built already. Models are evaluated by using function *predict()*, provided by R, and accessed through Rattle's Evaluate tab.

Once the model is developed and evaluated, the next task is to deploying the same. A simple approach to deployment in R is to use predict() to apply the

```
#Random Forest
student<- read.csv("G:/R_systems/Rmining/EDM_classifier.csv")
library(randomForest)
student[strain]
student[stVal,]
table(student[stTrain,]$grade)
table(student[stVal,]$grade)
formula <- grade_num ~ prev+att_num+internal_num+midterm+lab
stud_rf <- randomForest(formula, student[stTrain,],ntree=1, proximity=TRUE)
table(predict(stud_rf, student[stTrain,]), student[stTrain,]$grade_num)
table(predict(stud_rf,student[stVal,]), Student[stVal,]$grade_num)
```

Figure 4.17 R Implementation of random forest classifier.

```
    2   3   4   5   6
   10  17  55  44  14
 > table(student[stVal,]$grade)

    2   3   4   5   6
    1   8  26  19   6
 > table(predict(stud_rf, student[stTrain,]), student[stTrain,]$grade_num)

                              2   3   4   5   6
 2                            1   0   0   0   0
 2.30769230769231            9   2   1   0   0
 3                            0   5   0   0   0
 3.4                         0   9   3   0   0
 4                            0   0  32   2   0
 4.2                         0   0   5   3   0
 4.33333333333333            0   0   2   3   0
 4.5                         0   0   3   2   0
 4.6                         0   0   2   4   0
 5                            0   1   4  13   1
 5.25                        0   0   1   3   3
 5.33333333333333            0   0   0   6   2
 5.5                         0   0   2   7   4
 6                            0   0   0   1   4
 > table(predict(stud_rf, student[stVal,]), student[stVal,]$grade_num)

                              2   3   4   5   6
 2                            1   0   0   0   0
 2.30769230769231            0   0   1   0   0
 3                            0   2   0   0   0
 3.4                         0   3   1   0   0
 4                            0   2  13   3   1
 4.33333333333333            0   0   1   0   0
 4.5                         0   0   4   1   0
 5                            0   1   4   4   1
 5.25                        0   0   2   4   0
 5.33333333333333            0   0   0   2   2
 5.5                         0   0   0   4   0
 6                            0   0   0   1   2
```

Figure 4.18   Random forest classification results in R.

model to a new dataset. The scoring is achieved in Rattle by selecting *Score* option under *Evaluation* tab.

The performance of the classifiers for target variable "*grade*" can be described by the "confusion matrix", a squared contingency table. It compares predictions with actual answers. This also introduces the concepts of true positives, false positives, true negatives, and false negatives. The number of correctly classified records is the summation of diagonals in the matrix; all others are incorrectly classified. Since there are five classes the order of confusion matrix 5×5.

4.11.1 Model Evaluation in R

Evaluation of classifiers based on confusion matrix is explained in Table 4.1. Classification models are evaluated on the basis of their accuracy, that is, the

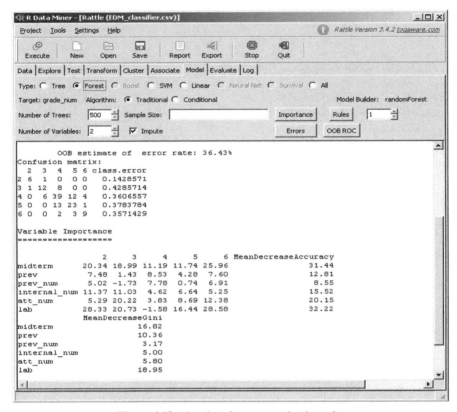

Figure 4.19   Random forest execution in rattle.

percentage of observations correctly classified. It is calculated as the ratio between the sum of the diagonal elements of the confusion matrix and the sample size. Method of evaluation revels that Naïve Bayes is the suitable model in this case is considered as classifier for the predication of student' academic performance.

Figure 4.22 shows, using scatter plot, the distribution of the performances for each of the four classifiers. This graph confirms the conclusion drawn from Table 4.1.

4.11.2 Model Evaluation in Rattle

Rattle's *Evaluate* tab gives access to variety of options for evaluating the performance of classifiers. This is listed in Figure 4.23. The series of different

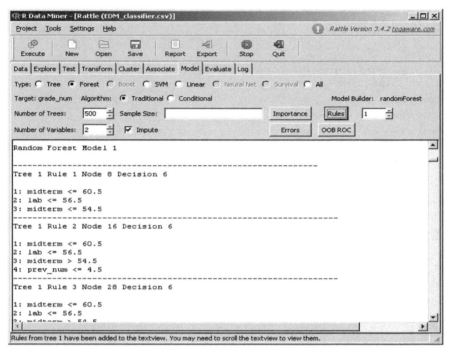

Figure 4.20   Rule set generated in random forest model.

Types of evaluations is presented as a set of radio buttons from Confusion Matrix to Score. At any time, only one type of evaluation is permitted. Next to row of evaluation types Rattle provides a set of check boxes to select the models wish to evaluate. These check boxes are enabled once a model has been built. Next is the option for dataset used for evaluation. The options here are *training*, *validation*, *test* and *full*. The testing dataset is the one, which is not at all used during model building.

The easiest measure of the performance of a model is the error rate. This is calculated as the percentage of observations for which the model wrongly predicts the class with respect to the actual class. This is done by simply divide the number of wrongly classified observations by the total number of observations.

Other types of measures are precision, sensitivity, and specificity. The *precision* of a classifier is the ratio of the number of true positives to the total number of predicted positives. The *sensitivity* of a model denotes the true positive rate. *Specificity* of a model is denotes the true negative rate.

Figure 4.24 shows confusion matrices for the built models and selected datasets. It gives matrix for decision tree model. Scrolling the window gives confusion matrices for other models.

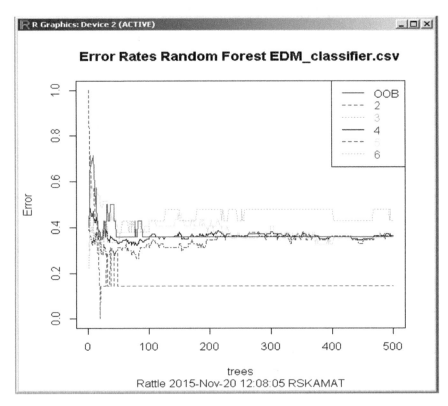

Figure 4.21   Error plot for random forest model.

Table 4.1   Classifiers evaluation metrics

Classifier	Expected Result	Confusion Matrix		Accuracy
Naïve Bayes	Training Set	Training Set	Validation Set	87.5%
	2 3 4 5 6	2 3 4 5 6	2 3 4 5 6	
	9 21 55 38 17	2 9 0 0 0 0	2 2 0 0 0 0	
		3 0 21 3 0 0	3 0 4 0 0 0	
	Validation Set	4 0 0 45 3 0	4 0 0 23 2 0	
	2 3 4 5 6	5 0 0 7 34 1	5 0 0 3 23 0	
	2 4 26 25 3	6 0 0 0 1 16	6 0 0 0 0 3	

(Continued)

Table 4.1   Continued

Classifier	Expected Result	Confusion Matrix		Accuracy

Neural Network	Training Set
	A B E O X
	55 44 10 17 14
	
	Validation Set
	A B E O X
	26 19 1 8 6
		Training Set
		   A B E O X
		A 46 7 0 2 0
		B 8 36 0 0 0
		E 0 0 10 0 0
		O 2 0 0 15 0
		X 1 2 0 0 11
			Validation Set
			   A B E O X
			A 14 5 0 3 4
			B 6 11 0 0 2
			E 0 0 1 0 0
			O 3 0 0 5 0
			X 1 4 0 0 1
				75%

Random Forest	Training Set
	2 3 4 5 6
	10 17 55 44 14
	
	Validation Set
	2 3 4 5 6
	1 8 26 19 6
		Training Set
		  2 3 4 5 6
		2 10 2 1 0 0
		3 0 14 3 0 0
		4 0 0 44 14 0
		5 0 1 7 29 10
		6 0 0 0 1 4
			Validation Set
			  2 3 4 5 6
			2 1 0 1 0 0
			3 0 5 1 0 0
			4 0 2 18 4 1
			5 0 1 6 14 3
			6 0 0 0 1 2
				70%

Decision Tree	Training Set
	2 3 4 5 6
	10 17 55 44 14
	
	Validation Set
	2 3 4 5 6
	1 8 26 19 6
		Training Set
		  2 3 4 5 6
		2 10 5 1 0 0
		3 7 14 3 0 0
		4 0 1 28 10 0
		5 0 0 12 31 3
		6 0 0 1 3 11
			Validation Set
			  2 3 4 5 6
			2 1 1 0 0 0
			3 0 5 4 0 0
			4 0 2 20 8 1
			5 0 0 2 10 4
			6 0 0 0 1 2
				66%

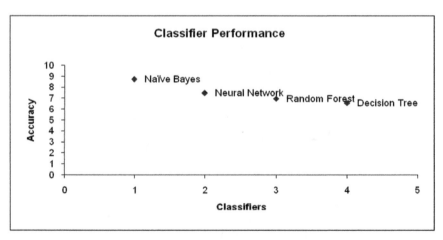

Figure 4.22   Scatter plot of the performance indices for classifiers.

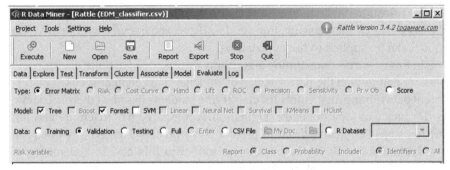

Figure 4.23   Evaluate tab options in rattle.

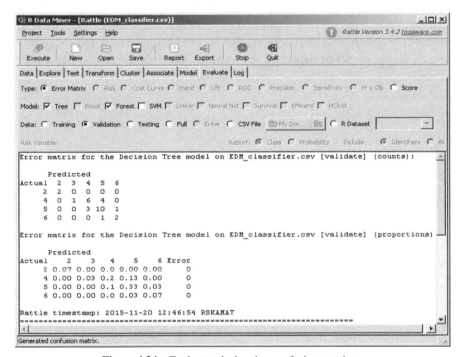

Figure 4.24   Evaluate tab showing confusion matrices.

4.12 Conclusion

Classification is one of the most regularly studied and utilized method by data mining and machine-learning researchers. It includes predicting the value of a variable based on the values of other variables. The classification process includes learning and testing. During learning, the training dataset is analyzed by classification model. During validation, test dataset is used to approximate the accuracy of the classification model. If the accuracy is satisfactory, the model can be applied to the new dataset.

In this research, data mining algorithms were utilized to build classifiers to classify and predict the performance of students. This chapter focused on different data mining algorithms for classifying students based on their academic performance in semester end examination. A comparison of machine-learning methods has been carried out to predict performance. Method of evaluation revels that Naïve Bayes is the suitable model in this case is considered as classifier for the predication of student' academic performance.

Classification is one of the powerful techniques in data mining to build models from an input dataset which can be used for predicting students' academic performance in current end of semester examination. It is used to model continuous-valued functions, that is, predicts unknown values. In this model, single aspect of data deduced from some combination of other aspect of data.

5

Educational Data Analysis: Clustering Approach

5.1 Introduction

Data mining techniques are used in the field of education to operate on large volume of data to discover hidden pattern and relationship helpful in decision-making. Clustering is finding groups of objects in such a way that the objects present in one group will be alike to one another and different from the objects in another group. Clustering a set of n objects into k groups is usually moved by the aim of identifying internally homogenous groups according to a specific set of variables. It can be said as identification of similar classes of objects. By using this technique, we can further identify dense and sparse regions in object space and can discover overall distribution pattern and correlations among data attributes.

Clustering can be considered the most important unsupervised learning technique. It is mainly useful in cases where the most common types within the dataset are not well known in advance. If a group of clusters is optimal, each data item in a cluster will in general be more alike to the other data items in that cluster than data items in other clusters. The objective of clustering was to find a structural representation of data by grouping similar data items together. A cluster has high similarity in association with one another but is very dissimilar to objects in other clusters. Classification approach can also be used for effective means of distinguishing groups or classes of object, but it becomes costly so clustering can be used as preprocessing approach for attribute subset selection and classification. It is a process which divides data into groups of similar objects.

From a machine-learning perspective, clusters correspond to hidden patterns and search for clusters is categorized as unsupervised learning. From a practical perspective, clustering plays an outstanding role in various data mining applications such as information retrieval and text mining, scientific data exploration, web analysis, spatial database applications, medical diagnostics, CRM, marketing, computational biology, and many others. Clustering algorithms aim at finding homogeneous groups in data. Various clustering techniques are used in this research and comparative study of these techniques is explained in this chapter.

5.2 Clustering in Educational Data Mining

In educational data mining, clustering can be used to group the students as per their activities. For example, distinguishing active students from less active students according to their performance can be done by clustering. According to clustering, clusters distinguish student's performance according to their behavior and activates.

Clustering analysis is a common unsupervised learning technique. Its aim was to group objects into different categories. That is, a collection of data objects that are similar to one another are grouped into the same cluster and the objects that are dissimilar are grouped into other clusters. The aim of unsupervised learning was to recognize patterns in the data that broaden our knowledge. There is no precise target variable that we are attempting to model. Instead, this approach enlightens on the patterns that emerge from the descriptive analytics.

In this approach, observations are grouped in an unguided fashion according to how similar they are. Grouping is done on the basis of a measure of the distance between observations. For example, the dataset referred in this book made up of students of different grades. Clustering approach groups this dataset into smaller, more definable groups. Observations within the group are close together but are quite separate from other groups. These clusters are useful in grouping observations to manage datasets easier.

Various algorithms have been developed for clustering. The most popular method for prediction in clustering is K-means. K-means clustering algorithm is used to automatically cluster the students. This method is used to classify the students' performance according to the learning style which is visual, active, and sequential.

5.3 Experimental Setup

The clustering algorithm in R and Rattle accepts the dataset in .csv format. The dataset consists of 200 instances with 6 different attributes. It accepts the nominal data and binary sets. Student dataset contains categorical values,

Table 5.1 Categorical variables "attendance" and "internals" numerical conversion

Attendance/Internals	Numerical Assignment
good	1
average	2
poor	3

which requires preprocessing for further process. As clustering algorithms works on numeric values only, categorical variables are converted in to numerical form. Categorical variables such as "attendance" and "internals" are converted to numerical form as explained in the Table 5.1. Partial dataset is shown in Figure 5.1.

5.4 Clustering Techniques

The objective of clustering was to find high-quality clusters such that the inter-cluster distances are maximized and the intra-cluster distances are minimized. In clustering, the aim was to find data items that naturally group together, dividing the entire dataset into a group of clusters. It is particularly useful when the most similar categories within the data base are not known in advance. If a group of clusters is finest, within a category, each data item will in common be more similar to the other data items in that cluster than data items in other clusters.

Clustering and its classification are shown in Figure 5.2. The hierarchical clustering functions basically in joining closest clusters until the desired number of clusters is achieved. This kind of hierarchical clustering is named agglomerative because it joins the clusters iteratively. There is also a divisive hierarchical clustering that does a reverse process, every data item begin in the same cluster, and then it is divided into smaller groups.

In clustering, the aim was to find data items that naturally group together, dividing the full dataset into a group of clusters. Clustering algorithms typically split into two categories:

- Hierarchical approaches such as hierarchical agglomerative clustering assume that clusters themselves cluster together.
- Non-hierarchical approaches such as K-means, Gaussian mixture modeling, and spectral clustering approaches assume that clusters are separate from each other.

Many clustering methods have been proposed and they can be broadly classified into four categories:

prev	att_num	internal_num	midterm	lab	sem
74	1	1	67	72	73
72	2	1	65	54	62
51	1	2	61	62	58
62	2	1	65	63	60
76	1	1	71	78	81
73	1	2	65	67	70
78	1	3	63	62	61
73	2	1	68	60	68
74	2	2	65	58	62
71	2	3	62	72	62
52	3	1	64	65	58
54	3	2	55	56	57
55	3	3	52	58	51
64	3	1	62	62	61
69	3	2	58	61	59
61	3	3	53	57	58
75	3	1	64	65	62
71	3	2	64	65	63
70	3	3	62	69	65
57	2	1	57	61	61
51	2	2	57	62	60
52	2	3	54	52	43
57	1	1	61	68	63

Figure 5.1   Preprocessed sample dataset.

- partitioning methods,
- hierarchical methods,
- density-based methods, and
- grid based methods.

Clustering is a descriptive and unsupervised learning task. Clustering is quite similar to classification except that the groups or classes are not predefined. In clustering, no parameter is selected as a target but the relationship between parameters can be discovered based on the formed clusters.

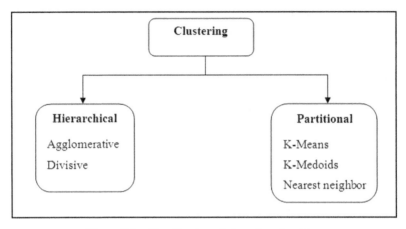

Figure 5.2 Classification of clustering algorithm.

5.5 Classification via Clustering—Design Framework

This research implements a meta-classifier that uses a cluster for classification approach based on the assumption that each cluster corresponds to a class. Figure 5.3 shows framework of classification via clustering. Firstly, the students' academic performance details have collected and preprocessed. Next, a clustering algorithm is executed using the training data, after removal of the class attribute, and the mapping between classes and clusters is determined. This mapping is then used to predict class labels for unseen instances in test data. In other words, the class attribute is not used in clustering, but it is used to evaluate the obtained clusters as classifiers.

For all cluster algorithms, it is important to ensure that the number of clusters generated is the same as the number of class labels in the dataset in

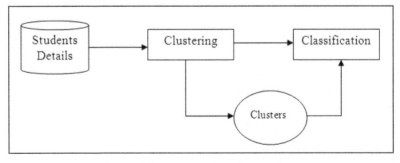

Figure 5.3 Classification via clustering.

order to obtain a useful model that relates each cluster with one class. This approach is to test student's academic performance in terms of grade.

5.6 Cluster Analysis in R and Rattle

R and Rattle, data mining tools allow the user to analyze data from different dimensions categorize it and summarize the relationship. Cluster analysis used to segment a large set of data into subsets. Each cluster is collection of data objects that are similar to another placed within the same cluster but dissimilar to objects in other cluster. Clustering is one of the basic techniques often used in analyzing datasets.

In educational data mining, clustering has been used to group the students according their behavior. For example, clustering can be used to distinguish clever students from dull students according to their performance in exam. Section 5.7 to 5.10 explains and implements different types of clustering in R and Rattle to segment students into groups according to their academic performance. Clustering can be considered the most important unsupervised learning technique.

5.7 K-means Clustering

The K-means algorithm identifies a collection of k clusters using a heuristic search starting with a selection of k randomly chosen clusters. Cluster analysis is based on measuring similarity between objects by computing the distance between each pair. The objective of this K-means test was to choose the best cluster center to be the centroid. The K-means algorithm requires the change of nominal attributes into numerical. The K-means algorithm takes the input parameter, k, and partitions a set of n data items into k clusters in such a way that the resulting intra cluster likeness is high but the inter cluster likeness is low. Cluster similarity is measured with respect to the mean value of the data items in a cluster, which can be viewed the cluster's centroid or center of gravity.

First, it randomly selects k of the data items, each of which in the beginning represents a cluster mean. For each of the remaining data items, a data item is assigned to the most similar cluster, based on the distance between the data item and the cluster mean. Then, it computes the new mean for each cluster. Same process iterated until the criterion function converges. Typically, the square-error criterion is used, defined as

$$E = \sum_{I=1}^{K} \sum P \in C_i \, |p - m_i|^2,$$

where E is the sum of the square error for all objects in the dataset; p is the point in data space representing a given data item; the mean of cluster C_i is m_i. In other words, for each object in each cluster, the distance from the object to its cluster center is squared, and the distances are summed. This principle tries to construct the resulting k clusters as compacted and as separate as possible.

In K-means clustering technique, clusters are wholly dependent on the choice of the initial cluster centroids. K data items are chosen as initial cluster centers followed by the distances of all data items are calculated by Euclidean distance formula. Data items having less distance to centroids are shifted to the appropriate cluster. This process is continued until no more alterations occur in clusters. The Figure 5.4 shows basic K-mean clustering algorithm steps.

INPUT: Number of desired clusters K

Students Records D= {d₁, d₂...dₙ}

OUTPUT: A set of K clusters

Steps:

1. Randomly select k data objects from data set D as initial centers.

2. Repeat;

    a. Calculate the distance between each data object d_i (1<= i<=n) and all k clusters Cj(1 <= j<-k) and assign data object d_i to the nearest cluster.

    b. Calculate for each cluster Cj(1 <= j<=k) recalculate the cluster center.

  Until no change in the center of clusters.

3. Time complexity of K-mean Clustering is represented by

       O(nkt)

Where n is the number of objects, k is the number of clusters and t is the number of iterations.

Figure 5.4   Steps of basic K-mean clustering algorithm.

5.7.1 K-means Clustering in R

The *stats* package must be installed in order to use the K-means algorithm in R. This package consists of a function that performs the K-Mean clustering, according to different algorithms such as *Llioyd, Forgy, MacQueen, Hartgan.* R clustering aims to group these student dataset into K=6 groups based on their attribute values. R script for implementation of K-means clustering is given in Figure 5.5.

The K-means process can be used to define the clusters as shown in Figure 5.6. The implementation of K-means generated five clusters, relatively homogeneous, consisting of 18, 47, 55, 34, and 46 tuples. Analyzing the cluster means, each group can be related with each of the five classes of student:

- First and second groups have higher cluster means for *previous sem, midterm,* and *sem* performance, where as fourth cluster has lower value for these
- Value of *attendance* performance high for second, third, and fourth clusters
- Intra cluster analysis revels that marks are in increasing order among the clusters, for example first cluster *previous sem* performance value higher than second cluster
- First and fifth clusters current semester performance value more than *previous sem* performance

5.7.2 K-means Clustering in Rattle

The *student* dataset is used to illustrate the building of cluster model in Rattle. As clustering works on nominal data only, the preprocessed nominal dataset is loaded into Rattle by selecting *data* tab. Corresponding execution is shown in the Figure 5.7.

```
#kmeans clustering
student<- read.csv("G:/R_systems/Rmining/EDM_clustering.csv")
library(stats)
km <- kmeans(student, 5, iter.max = 15)
print(km)
plot(student, col = student$cluster)
points(student$centers, col = 1:2, pch = 8)
```

Figure 5.5   K-means clustering implementation in R.

```
> km <- kmeans(student, 5, iter.max = 15)
> print(km)
K-means clustering with 5 clusters of sizes 18, 47, 55, 34, 46

Cluster means:
      prev  att_num internal_num midterm      lab      sem
1 77.83333 1.333333     1.333333 78.44444 78.44444 79.27778
2 72.02128 2.191489     2.063830 64.31915 61.72340 61.08511
3 58.85455 2.181818     1.800000 59.32727 61.43636 60.58182
4 54.05882 2.029412     2.411765 54.20588 52.50000 50.52941
5 68.06522 1.521739     1.586957 67.36957 69.21739 69.50000

Clustering vector:
  [1] 5 2 3 3 1 5 2 2 2 5 3 4 4 3 2 3 2 2 5 3 3 4 3 3 4 5 3 2 5 2 2 5 1 1 5 5 5 2 3 4 4 5 3 3 5
 [46] 2 2 3 4 4 3 4 4 5 3 3 5 5 2 3 5 2 5 2 5 2 5 2 4 4 4 3 2 3 2 2 2 3 4 3 5 4 4 3 3 3 5 5 5 1 1 1
 [91] 1 1 2 5 5 1 4 5 4 3 5 2 3 2 1 5 2 2 2 2 3 3 4 2 3 3 2 2 5 3 3 4 3 3 3 3 5 2 5 5 2 5 1 1 5
[136] 5 5 2 3 4 3 3 3 3 5 2 3 3 4 4 3 4 4 3 3 3 5 5 2 3 5 2 5 2 5 2 4 4 4 2 3 3 2 2 2 3 3 4 5 4
[181] 4 3 3 4 5 2 2 1 1 1 1 2 2 5 5 1 4 5 4 1

Within cluster sum of squares by cluster:
[1] 1593.000 2790.340 3048.836 1935.618 3845.478
 (between_SS / total_SS =  76.5 %)
```

Figure 5.6   Output of K-means clustering implementation in R.

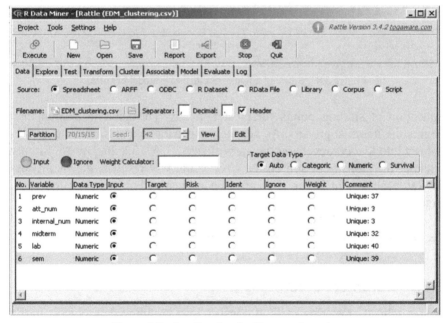

Figure 5.7   Loading data for K-means in rattle.

The *Cluster* tab in the Rattle window gives access to various clustering techniques, including K-means. *K-means* is provided directly through R by the standard *stats* package. It is the default option to build cluster model. A random seed is provided. Changing the seed will result in an arbitrarily different

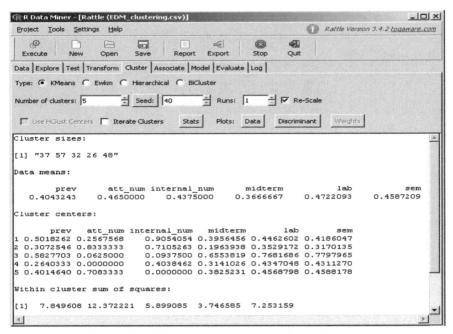

Figure 5.8   Building a K-means clustering model in rattle.

collection of starting points for cluster means. The heuristic search then begins the iterative process. As per the input parameters, Rattle categorizes dataset into five homogeneous clusters of size 37, 57, 32, 26, and 48. Simply clicking the *Execute* button on the *Cluster* tab will result K-means process in Rattle shown in Figure 5.8 and the corresponding plotting shown in Figure 5.9.

The text view lists cluster means. K-means clustering consists of five vectors of the mean values for each of the variables.

5.8 Hierarchical Clustering

A hierarchical clustering method consists into grouping data objects into a tree of clusters. The metrics used to merge or split clusters different distance measures. There are two main types in hierarchical clustering:

- Bottom-up approach starts with small clusters composed by a single object and, at each step; merge the current clusters into greater ones,

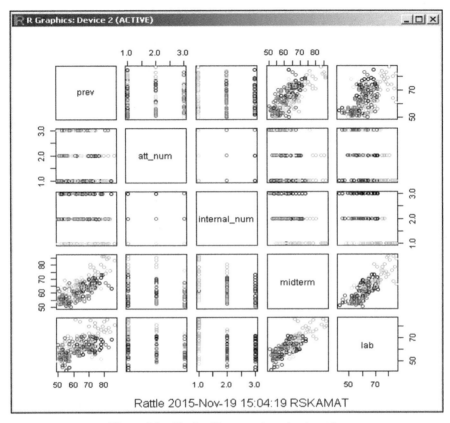

Figure 5.9 Plotting K-means clustering in rattle.

successively, until reach a cluster composed by all data objects. R function *diana* follows bottom-up approach.

- Top-down approach uses the same logic, but to the opposite direction, starting with the greatest cluster, composed by all objects, and split it successively into smaller clusters until reach the singleton groups. R function *agnes* follows bottom-up approach.

5.8.1 Hierarchical Clustering in R

R implementation of top-down hierarchical clustering is given in Figure 5.10. The groups defined by hierarchical clustering are shown in Figure 5.11.

```
# hierarchical clustering
student<- read.csv("G:/R_systems/Rmining/EDM_clustering.csv")
library(cluster)
ag <- agnes (student, 0, metric="euclidean", 0, method ="single")
print(ag)
plot(ag, ask = FALSE, which.plots = NULL)
```

Figure 5.10   Hierarchical clustering implementation in R.

```
> student<- read.csv("G:/R_systems/Rmining/EDM_clustering.csv")
> library(cluster)
> ag <- agnes (student, 0, metric="euclidean", 0, method ="single")
> print(ag)
Call:    agnes(x = student, diss = 0, metric = "euclidean", stand = 0,       method = "single")
Agglomerative coefficient:  0.6900321
Order of objects:
  [1]    1    2    3   21   39  125   97  103   25    4   14   70   82  156  171   55  155  183
 [19]  144   83  143  170   16   43  120  141   72  147  131   17   18   74   66  162  107  166
 [37]   31   59  186   71  174  192   62   64   73  118  173   28  159  128  164   84   30   86
 [55]  104  114   87   42   54   12   40   53   52   80  153  180  124  149   49   78  172  154
 [73]   38   24   15  115   56    9  100   32   61  161   65  165   27   20   60  139  160  148
 [91]   48   23  151   51  111    8  138  132   47   46   26  127   19   57   85  195  185   95
[109]   98   58  157  158  198   13   41   11   10  146  140  126  123  110  109  108  101   45
[127]  119  145   44  130  142  182   77  177   76  116    7  135  113  112   22   50  152   99
[145]  199  150   69  169  178  122   67  197  167   68  168    6   36  106   37  136  137  121
[163]   81  181  176   35  163   63   93  193  102   79  179   34  134    5   33  105  133   96
[181]  196   89  189   91  191   92  187   94  194   75  175  117  184   88  188  200   90  190
[199]   29  129
Height (summary):
   Min. 1st Qu.  Median   Mean 3rd Qu.    Max.
  0.000   3.162   3.873   3.944   4.527  11.270

Available components:
[1] "order"  "height"  "ac"        "merge"   "diss"     "call"      "method" "data"
```

Figure 5.11   Output of hierarchical clustering implementation in R.

5.8.2 Hierarchical Clustering in Rattle

Figure 5.12 gives details of hierarchical clustering executed in Rattle. Five clusters of homogeneous in nature designed with *Euclidean* distance and *ward* as agglomerate parameter. Refer figure for additional details. Result in terms of cluster density depicted as data plot and discriminant plot in Figures 5.13 and 5.14, respectively.

After building the model in Rattle, the Stats, Data Plot, and Discriminant Plot buttons become available.

- Stats button result in quite a few additional cluster statistics being displayed in the text view. These statistics helps in determining the quality of the model.
- The execution of Data Plot and the Discriminant Plot buttons result in plots that exhibit how the clusters are distributed across the dataset.

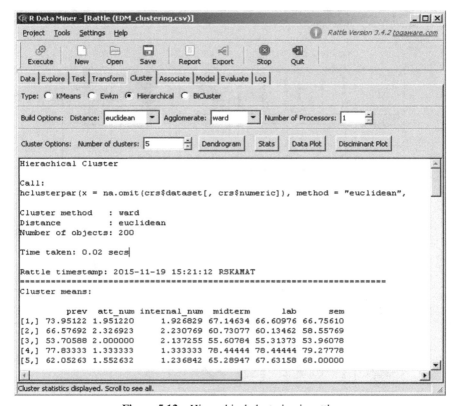

Figure 5.12 Hierarchical clustering in rattle.

5.9 Hybrid Hierarchical Clustering

Hybrid hierarchical clustering techniques try to combine the best advantages of both agglomerative and divisive techniques. Agglomerative algorithms group data into many small clusters and few large ones, which usually makes them good at identifying small clusters but not large ones. Divisive algorithms, however, have reserved characteristics; making them good at identifying large clusters in general.

This section presents hybrid hierarchical algorithm that uses the concept of "*mutual cluster*" to combine the divisive techniques procedures with information gained from a preliminary agglomerative clustering. A mutual cluster can be defined as a group of data that are collectively closer to each other than to any other data, and distant from all other data. Implementation of hybrid clustering in R given in Figure 5.15.

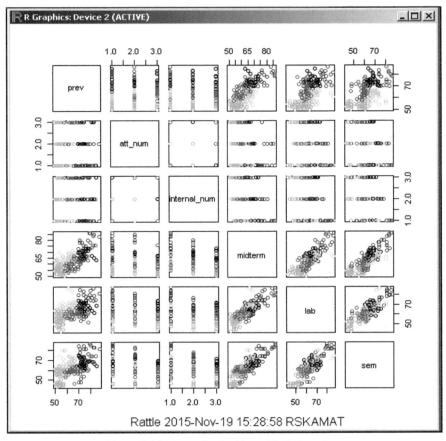

Figure 5.13   Data plot for hierarchical clustering.

The hybrid hierarchical algorithm using mutual clusters can be described in three steps:

1. Compute the mutual clusters using an agglomerative technique.
2. Perform a constrained divisive technique in which each mutual cluster must stay intact. This is accomplished by temporarily replacing a mutual cluster of data elements by their centroid.
3. Once the divisive clustering is complete, divide each mutual cluster by performing another divisive clustering "within" each mutual cluster.

Corresponding dendogram for the dataset is shown in Figure 5.16. It depicts grouping of tuples in hybrid clustering. Because the case study has a

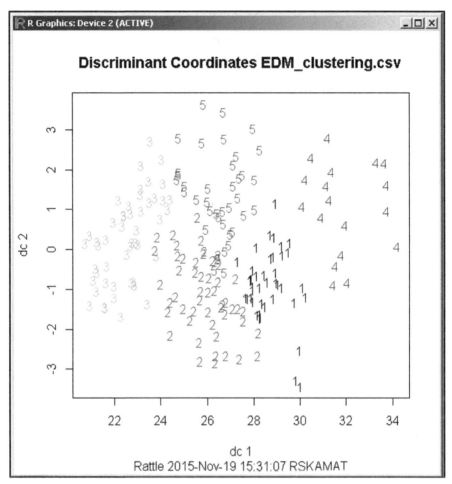

Figure 5.14 Discriminant plot for hierarchical clustering.

multi-dimensional data it's harder to visualize what the clusters have in common. However, when a good number of clusters are chosen, it is possible to see their similarities and better analyze the results.

5.10 Density-based Clustering

As name indicates, this technique is based on density for deciding which clusters each element will be in. The idea behind constructing clusters based on the density properties of the database is derived from a human natural

```
# Read the data file
student<- read.csv("G:/R_systems/Rmining/EDM_clustering.csv")
# calculate the mutual clusters
mc <- mutualCluster(student)
# Calculate the Hybrid Hierarchical Cluster
hyb <- hybridHclust(student, mc)
# Plot the Hybrid Clustering Dendogram
plot(hyb)
```

Figure 5.15   Hybrid clustering implementation in R.

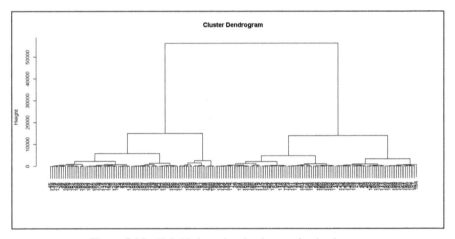

Figure 5.16   Hybrid clustering dendogram for the dataset.

clustering approach. The clusters and consequently the classes are easily and readily identifiable because they have an increased density with respect to the points they possess. On the other hand, the single points scattered around the database are outliers, which means they do not belong to any clusters as a result of being in an area with relatively low concentration.

R function *dbscan* is used for density-based clustering in this research. This function is available in *fpc* package. This algorithm requires at most two parameters: a density metric and the minimum size of a cluster. As a result, estimating the number of clusters a priori is not a need, as opposed to other techniques, namely K-means. Figure 5.17 gives R implementation of density-based clustering. *dbscan* function is called with eps, reachability distance is 10, and MinPts is set to its default, which is 5. Figure 5.18 shows the results after the clustering has been performed by the DBSCAN algorithm.

```
library(fpc)
student<- read.csv("G:/R_systems/Rmining/EDM_dbscan.csv")
d <- dbscan(student,10,showplot = 2);
```

Figure 5.17 Density-based clustering implementation in R.

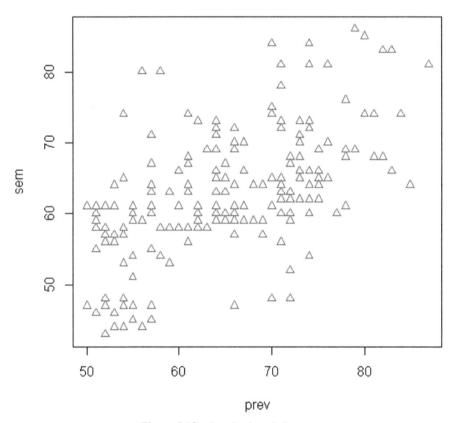

Figure 5.18 Density-based clusters.

5.11 Conclusion

The basic tuning option for creating cluster models in Rattle is simply the number of clusters that are to be created. The default is 10, but any positive integer greater than 1 is allowed. The Seed option permits different starting points to be used for the heuristic search. A different seed value usually results into different model.

Educational data analysis using clustering and design of prominent clusters have been explained in this chapter. Clustering is finding groups of data items such that the objects in one cluster will be similar to each other and different from the objects in another group. Thus, it allows the data miner to break dataset into more significant groups and then contrast the dissimilar clusters against each other.

From a machine-learning perspective, clusters correspond to hidden patterns and search for clusters is categorized as unsupervised learning. This chapter focused on implementation of unsupervised machine-learning algorithms for clustering derived from the R data mining tool.

6

Epilogue and Further Directions

6.1 A Gist of Final Setup

Mining of Educational Data for the Analysis and Prediction of Students' Performance intended to be done during the research has finally fulfilled. The primary objective of this research was to use data mining techniques to study and analyze student's academic performance. This research is a case study of an academic institute to improve the quality of education by analyzing the data and discover the factors that affect the academic results. It focused on the implementation of data mining techniques and methods for acquiring new knowledge from student database.

Incorporation of data mining applications in educational systems carried out in this research. In supervised learning phase, classifier models are developed which associates the class attribute and the explanatory attributes using a learning set randomly selected from the dataset. Performance of the each classifier evaluated and selects the "best" one. It allows checking the performance of the trained classification models against a test set, evaluating their predictive correctness with data not used in the learning step. In unsupervised learning phase, various clustering techniques are applied to get prominent clusters. EDM research here focused with developing new methods to find out knowledge from educational database. It has proved how useful data mining can be in higher education, particularly to improve students' performance.

6.2 Research Findings

The main goal of the research was to reveal the high potential of data mining applications for students' performance management. This research justifies the abilities of data mining techniques in perspective of higher education by providing a data mining model for education system. The full integration of data mining in the educational environment will become a reality, and fully operative implementations could be made available not only for researchers and developers but also for external users. This research is a case study of EDM process for analysis and prediction of students' academic performance using the R/Rattle data mining tool and selected R packages for computing.

6.2.1 Role of EDM

In recent years, educational organizations are facing the biggest challenge that the explosive growth of data and to usage of this data for the improvement of managerial decisions' quality. Educational data mining is an interesting research area which extracts useful, previously unknown patterns from educational database for better understanding, improved educational performance, and assessment of the student learning process. Educational dataset contains the helpful information for predicting students' performance, rank factor and details. Educational data mining is concerned with implementing methods for discovering knowledge from dataset and also helpful in classifying educational database. The data mining prediction technique supports decision-making which can assist for students' performance.

6.2.2 Experimental Results

This research presented the potential use of education data mining using classification and clustering techniques in enhancing the quality and predicting students' performances. The student evaluation factors are studied and analyzed for this purpose. This study will help to identify the students who need special coaching and allow the teacher to provide proper advising. This research has collected real-time data that describing the relationships between learning behavior of students and their academic performance. It has accessed the related databases of CSIBER containing a set of 200 records of Master of Computer Application students. After preprocessing the data, research applied data mining techniques in R and Rattle for the analysis of educational data.

A couple of EDM methods and algorithms are adopted in this research to discover hidden patterns and relationships. In each of these tasks, research has presented the extracted knowledge and described its importance in educational domain. The prediction model is developed to predict the performance of students in semester examination. Classification, which is one of the prediction types classified the data, based on the training set and uses the pattern to classify a new data. Clustering is applied to group the records in classes that are similar, and dissimilar records to other classes.

6.2.3 Student Segmentation Using Clustering

From a machine-learning point of view, clusters correspond to hidden patterns and search for clusters is categorized as unsupervised learning. Its aim is to group objects into different categories. That is, a collection of data objects that are similar to one another are grouped into the same cluster and the objects that are dissimilar are grouped into other clusters. Screenshot of K-means clustering in R is shown in Figure 6.1. Various clustering techniques in R are used in this research and comparative study of these techniques is explained. Clustering aimed at finding high-quality clusters such that the inter-cluster distances are maximized and the intra-cluster distances are minimized.

Cluster mean of "current semester result" (sixth column of the Figure 6.1) is considered for analysis. Corresponding details is given in Table 6.1. Since the cluster centers of clusters 1 and 3 are closer treated as single. Thus, student dataset categorized as given in the following Table 6.2.

6.2.4 Classifiers to Predict Student Performance

Classification is one of the most frequently studied problems by data mining and machine-learning researchers. It consists of predicting the value of an attribute based on the values of other attributes. In this research, different

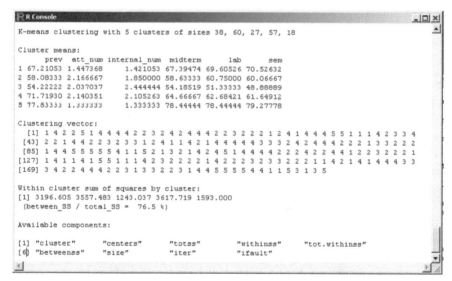

Figure 6.1 Screenshot of K-means clustering in R.

Table 6.1   Cluster analysis

Clusters	Cluster Center	Cluster Size
Cluster 0	70.52	38
Cluster 1	60.06	60
Cluster 2	48.88	27
Cluster 3	61.64	57
Cluster 4	79.27	18

Table 6.2   Categories of students based on K-means

Cluster Center	Cluster Size	Category
79.27	18	Fast Learners
70.52	38	Good Learners
60.85	117	Average Learners
48.88	27	Slow Learners

classification algorithms in R mining were used to build a classification model to classify and predict the performance of students. After developing classifiers, the criteria defined for evaluating their performances. Classification models are evaluated on the basis of their accuracy calculated using confusion matrices. Method of evaluation revels that Naïve Bayes is the suitable model in this case is considered as classifier for the predication of student' academic performance.

6.3 Conclusion

This book portrays "Educational Data Mining with R and Rattle". The present research carried out a data mining techniques implementation using different R packages. The machine-learning algorithms implemented are powered by a student dataset. Different classifiers are executed on student dataset. Naive Bayes is the clear winner in this contest. The results achieved through the basic set of variables described in Table 4.1 are encouraging, able to of execute supervised learning for classification projects accurately.

As we know large amount of data is maintained in educational database, so as to get required data and to discover the hidden relationship, various data mining techniques are used. Classification, clustering, outlier detection, association rule, prediction etc are varieties of popular data mining task within the educational data mining. This research is an application of state of art of EDM in analysis and prediction of students' academic performance.

6.3.1 Objectives Revisited

This research justified the abilities of data mining techniques in perspective of higher education by offering classifiers and prominent clusters for education system. Experiments have carried out in order to evaluate the performance and usefulness of different classification algorithms for predicting students' final marks based on information in the students' usage data.

The following objectives specified in the problem statement are fulfilled:

- Various classifiers have built and evaluated that tests certain attributes that may affect students' academic performance.
- Students' academic performance is forecasted based on their previous semester result, performance in class test, assignment and attendance.
- Model helped in identifying the students who need special counseling by the teacher.
- Set prominent clusters from student dataset designed and analyzed.

6.3.2 Insight of EDM

Educational data mining is a promising discipline, concerned with designing methods for exploring the exclusive types of data that come from educational institutes, and using those methods to better recognize students, and the environment in which they learn. New approaches derived from machine learning, statistics, scientific computing, psychometrics, etc helps in achieving this goal. EDM process converts raw data coming from educational systems into useful information that could potentially have a great impact on educational research and practice. There are various EDM methods and algorithms used to discover hidden patterns and relationships which include prediction, clustering, and relationship mining.

6.4 Recommendation for Future Research

For future work, a way to generalize the study to more diverse courses to get more accurate results needs to be developed. Present research dealt with building classifiers and the design of prominent clusters. As research goals moving forward, applying association rule mining, text mining. Data mining algorithms could be embedded into Moodle, e-learning system so that one using the system can be benefited from the data mining techniques.

References

[1] A.A. Aziz, N.H. Ismail, F. Ahmad, "Mining Students' Academic Performance", Journal of Theoretical and Applied Information Technology 53(3), 485–495, 2013.

[2] Abeer Badr El Din Ahmed, Ibrahim Sayed Elaraby, "Data Mining: A prediction for Student's Performance Using Classification Method", World Journal of Computer Application and Technology 2(2), 43–47, 2014.

[3] Brijesh Kumar Bhardwaj, Saurabh Pal, "Data Mining: A prediction for performance improvement using classification", International Journal of Computer Science and Information Security 9(4), April 2011.

[4] Pimpa Cheewaprakobkit. Study of Factors Analysis Affecting Academic Achievement of Undergraduate Students in International Program, 2013.

[5] Cristobal Romero, Sebastian Ventura, Pedro G. Espejo and Cesar Hervas, "Data Mining Algorithms to Classify Students".

[6] Dhara Patel, Ruchi Modi, Ketan Sarvakar, "A Comparative Study of Clustering Data Mining: Techniques and Research Challenges", IJLTEMAS, Volume III, Issue IX, September 2014.

[7] Graham J. Williams, Rattle: A Data Mining GUI for R, The R Journal 1/2, 45–54, 2009.

[8] Jayshree Jha, Leena Ragha, "Educational Data Mining using Improved Apriori Algorithm", International Journal of Information and Computation Technology, 3(5), 411–418, 2013.

[9] Matthew Berland, Ryan S. Baker, Paulo Blikstein, "Educational Data Mining and Learning Analytics: Applications to Constructionist Research", Springer, Dordrecht, 2014.

[10] Otobo Firstman Noah, Baah Barida, Taylor Onate Egerton, "Evaluation of Student Performance Using Data Mining Over a Given Data Space", International Journal of Recent Technology and Engineering 2(4), 101–104, 2013.

[11] Pal A. Kumar and S. Pal, Classification Model of Prediction for Placement of Students 2013.

[12] Paul R. Cohen and Carole R. Beal, "Temporal Data Mining for Educational Applications", International Journal of Software and Informatics, 3(1), 29–44, 2009.

[13] Prabhat Kumar, Berkin Ozisikyilmaz, Wei-Keng Liao, Gokhan Memik, Alok Choudhary, "High Performance Data Mining Using R on Heterogeneous Platforms", 2011 IEEE International Parallel & Distributed Processing Symposium.

[14] Pratiyush Guleria and Manu Sood, "Data Mining in Education: A Review on the Knowledege Discovery Perspective", International Journal of Data Mining & Knowledge Management Process, 4(5), 2014.

[15] Sadiq Hussain, G.C. Hazarika, "Educational Data Mining Model Using Rattle", International Journal of Advanced Computer Science and Applications 5(6), 22–27, 2014.

[16] Srecko Natek, Moti Zwilling, "Data Mining for Small Student Data Set – Knowledge Management System for Higher Education Teachers", International Conference on Management, Knowledge and Learning, 19–21 June, 2013, Zadar, Croatia, 1379–1389.

[17] Sonali Agarwal, G. N. Pandey, M. D. Tiwari, "Data Mining in Education: Data Classification and Decision Tree Approach", International Journal of e-Education, e-Business, e-Management and e-Learning 2(2), 2012.

[18] Suchita Borkar, K. Rajeswari, "Predicting Students Academic Performance Using Education Data Mining", International Journal of Computer Science and Mobile Computing, IJCSMC 2(7), 273–279, 2013.

[19] Sunita B. Aher, L.M. Lobo, "Data Mining in Educational System using WEKA", International Conference on Emerging Technology Trends (ICETT) 2011, Proceedings published by International Journal of Computer Applications® (IJCA).

[20] Syeda Farha Shazmeen, Mirza Mustafa Ali Baig, M.Reena Pawar, "Performance Evaluation of Different Data Mining Classification Algorithm and Predictive Analysis", IOSR Journal of Computer Engineering, 10(6), 01–06, 2013.

[21] Varun Kumar, Anupama Chadha, "Mining Association Rules in Student's Assessment Data", IJCSI International Journal of Computer Science Issues, 9(5), 211–216, 2012.

[22] V. Ramesh, P. Parkavi, P. Yasodha, "Performance Analysis of Data Mining Techniques for Placement Chance Prediction", International Journal of Scientific & Engineering Research 2(8), 2011.

[23] M.I. López, J.M. Luna, C. Romero, S. Ventura, "Classification via clustering for predicting final marks based on student participation in forums".

[24] Yanchang Zhao, R and Data Mining: Examples and Case Studies.

[25] http://www.rdatamining.com

[26] Graham Williams, "Data Mining with Rattle and R: The Art of Excavating Data for Knowledge Discovery", Springer, Berlin.

[27] Luis Torgo, "Data Mining with R Learning with Case Studies", Taylor & Francis, New York.

Index

103

About the Authors

R. S. Kamath is Associate Professor in the Department of Computer Studies, Chhatrapati Shahu Institute of Business Education and Research, Kolhapur, India. She obtained her Bachelors and Masters in Computer Science from Mangalore University. She received her Ph.D. in Computer Science specialized in Computer Based Visualization from Shivaji University and completed the same in 2011. Dr. Kamath has to her credit 20 research papers published in reputed national and international journals and presented 11 papers in national conferences. She has completed two minor research funded by UGC. She is the author of two books and edited one seminar proceedings. The chapter entitled "Cost Effective 3D Stereo Visualization for Creative Learning – Virtual Reality in Education" is accepted for Encyclopedia of Information Science and Technology 4th Edition by IGI Global Publishing. Her areas of research interests are Artificial Intelligence, Virtual Reality, Soft Computing and Data Mining. She has immense skill of around twelve years in teaching and research.

R. K. Kamat holds the position of Professor in the Department of Electronics and heads the Department of Computer Science of Shivaji University, Kolhapur. He is Director of Internal Quality Assurance (IQAC) of the Shivaji University, Kolhapur. He obtained his Bachalors, Masters and M.Phil in Electronics from the Shivaji University, Kolhapur. Dr. Kamat gained his Ph.D. specialized in Smart Sensors from Goa University, Goa. Professor Kamat has published over 70 plus papers in International journals of repute and presented equal number of papers at National and International Conferences. He has published 10 books though reputed publishing house such as Springer UK. He has published scholarly literature on the quality issues in higher education. Five students have been awarded Ph.D. under his guidance and 11 more are working for their Doctorate. He has been involved with various initiatives of the apex organization at international level such as IEEE USA and Engineering Education group of Central Quinsland Australia. Through these organizations he is playing key role in spreading the scholastic culture by organizing conferences at various international destinations. Through this network he has visited various countries. Dr. Kamat is a recipient of the Young Scientist Award under the fast track scheme of the Department of Science and Technology (DST) of Government of India.